To B[...]

have [...]
christmas.

December 1997.

love
Graham.

LIFE ON MARS?

THE CASE FOR A COSMIC HERITAGE

LIFE ON MARS?

THE CASE FOR A COSMIC HERITAGE

Fred Hoyle and Chandra Wickramasinghe

Series editor and foreword: Paul R. Goddard

Published by:
Clinical Press Limited.
Registered Office:
Redland Green Farm
Redland
Bristol, BS6 7HF
UK

Hoyle, F. and Wickramasinghe, C.

Life on Mars? The case for a cosmic heritage.

ISBN: 1 85457 041 2

CONTENTS

FOREWORD

FOREWORD

By Paul Goddard, MD, FRCR

An amazing discovery

Is the Universe outside the Earth sterile or is it teeming with life? Did life originate on Earth or is it a cosmic event? Are there bugs in space, little green men on Mars and myriads of extraterrestrials, or are we completely alone on our single planet?

These and other age old debates about life in the Universe have been rekindled by the possible evidence of life on Mars (e.g. New Scientist, August 17th, 1996). But the fire never really died down and the two scientists who have kept the flame alight since the early 60s are Fred Hoyle and Chandra Wickramsinghe.

Imagine yourself in the situation that they found themselves. Highly respected astronomers researching on interstellar gas, they had already notched up many notable discoveries following on from Fred Hoyle's original, and subsequently proven correct, conclusion that there would be complex molecules in space. Trying to understand the spectra or the interstellar gas, for example from sources near the galactic centre (GC-IRS7), they discovered that the best fit was with organic material. More specifically an almost perfect correlation was obtained between the transmittance spectra of dessicated bacteria and the radiation flux from sources of cosmic infrared radiation.

Many people had previously postulated that life could be a cosmic phenomenon but before Hoyle and Wickramsinghe's discovery there was no objective evidence for the existence of life anywhere but on the Earth. Their evidence of organic life in space was thus a major discovery.

Further evidence has subsequently been forthcoming but the prevailing scientific theories have been strongly against the idea that life could have originated elsewhere. After all, the biologists may say, we know that life originated on earth and that it evolved due to natural selection by survival of the fittest. In such a scenario there is no need, and indeed no room, for a cosmic theory of life.

Thus many of the findings of Hoyle and Wickramsinghe have been
ridiculed although no cogent evidence to the contrary has ever been
put forward.

Now that the theory of abundant life in the cosmos is again
becoming worthy of respectable discussion, their fundamentally
important contributions are in danger of being passed over by scientists who would prefer to think that their own discoveries are all
new, and who would like to merge the new facts into their existing
neo-Darwinian culture.

But are the Darwinists really in such a strong position? Can
they really state as fact that the Darwinian theory of survival of the
fittest is the major determinant of biological life? A few moments
logical speculation is worthwhile at this point as to whether or not
survival of the fittest is the only significant determinant of life and
whether the biologists have such a strong case that the objective
evidence of the astronomers can simply be derided.

Do the fittest survive?

It is inherent in Dawinian and neo-Darwinian evolutionary theory
that the fittest survive and that this is the driving force of evolution
either at the level of the species, the individual or at the level of the
gene.

This thinking is also apparent in economics from the time of
Adam Smith onwards leading to the belief that the 'market knows
best'. It is instructive to compare biology and economics and to
wonder whether or not the commonality of ideas is simply a demonstration of the prevalence of ideas in society . . . ideas that have been
formulated together such that we may be programmed to accept
erroneous conclusions as fact.

The belief that the 'market knows best' has led to the assumption that left to its own devices "the market" will eventually promote the best products, the best techniques and the best possible
results. A disinclination to disagree with 'the market' is embodied
in the phrase of the marketeers "you can't buck the market".

So, we are led to believe, there is a similarity of evolution in nature to that in economics. They are leading to 'that which is better'.

I would suggest that is not a view that any respectable biologists or economist should now hold.

Is there evidence for classical evolution?

Evolution of species is a difficult subject. Although natural selection in species can occur and has been shown to occur in rare circumstances, it is more fundamentally challenging to note that the fossil evidence is lacking for the vast majority of the proposed evolution by natural selection. Indeed the fossil evidence would suggest that changes in biology have much more frequently occurred by jumps than by continuous change. Moreoever, the survivors are not necessarily the fittest for the habitat in which they find themselves.

Surely the reply is put forward for biology and economics as if it is conclusive: . . . they must be the fittest since they are the species/company that survived . . . they are therefore the fittest for survival? This argument is specious tautology.

Catastrophe

Consider the chairs around your dining table. They are used for sitting on and perhaps, for a variety of other uses. They may vary in their 'fitness' for the tasks set them. A bolt of lightning sets fire to your house and all the chairs on one side are burnt to a cinder but those on the other side are not destroyed because they are in a different place. Will the best chairs survive? Will the 'fittest' chairs be picked out? Maybe and maybe not! It would be a foolish person who assumed that after the catastrophic event the chairs that survived were the best for the job. By sheer bad luck it is quite likely that the best chair would be destroyed and the most rickety one survived unscathed!

Take a more realistic example. Two species of mammals could have evolved on two separate land masses. The ecology of both land masses could be almost identical and it could be that species A is a better animal than species B, could easily out-compete it for food and breeding sites, and is less harmful to its surroundings. Species A would replace B but for the fact that it has never come into contact with it due to the separation by a large body of water. The fittest mammal, we would agree, is species A.

Then the sea level rises by 20 metres. The land mass that species A is living on is all lower than 20 metres and all of the species A is wiped out. The land mass that B is living on has some areas that are above 20 metres and species B survives. At a later time the sea recedes and B is now able to colonise the land mass previously occupied by species A. Has the fittest survived? Catastrophe has led to the survival of the unfit.

The first fish in the pool may be a selfish fish

With economics as with biology, although catastrophe is clearly important, it is not essential for the survival of the unfit. For example the fittest creature may 'evolve' after the 'less-fit' but never get a chance to develop due to the larger number of 'less-fit' creatures that already occupy the particular ecological niche. This can be considered as the 'first fish in the pool' scenario.

Or in economics a number of different "less-fit" companies could co-operate to remove a better competitor. Consider in this context Betamax and VHS video systems or the various computer companies and their products. Clearly it is not always the best and fittest that survived.

The entire classical economic theory relies on a belief in 'enlightened self-interest'. If human beings are considered as rational creatures who will eventually choose the better product since it is 'right to do so', the economy will lead to stability and obey classical economic laws. In fact such stability never occurs and the products and companies that survive are not chosen solely for their fitness, although it is a factor, but for a whole host of reasons that appeal to the self-motivated and selfish individuals who make up the market (including, of course, myself!). The graph of the share index fluctuates in a manner that is mathematically chaotic rather than truly random, and does not appear at any time to be approaching 'sustainable' stability.

Again, in biology, if we ignore the selfishness of the individual, either at gene level, phenotype level, family group or species level we are binding ourselves to significant facts.

Is the system closed or open?

Another major consideration that we should address is whether or not we are dealing with a closed system. In a closed system it may make some sense to suggest that there will be a continual slide toward dissolution and entropy. Thus the greatest complexity will have occurred initially and later increases of complexity in a species (individual/gene/company/product) can only occur at the expense of other individuals. Thus there is an assumption that the survival of the fittest occurs at the expense of the other species or companies and that this is the way of the world . . . some must fall by the way-side in order that others may 'progress'.

Inherent in this belief is the assumption that at some mythical time there was much greater variety than now. Thus a choice could be made between the fit and the unfit leaving only the fit to survive. This is another assumption that has become enshrined in our belief system. Put simply we are saying that there must initially be a huge variety of possibilities and some have to fail in order that others survive.

But perhaps this is also untrue! Perhaps the whole basis of the argument is flawed and there is a place for all of the species or companies. This could only be the case if we are not living and existing in a closed system, and at the most basic level it is evident that we are not!

The Earth is not a closed system. It is bombarded with electro-magnetic radiation from the Sun which provides the vast proportion of the energy required for living organisms. Other irradiation comes from far distant sources as cosmic rays resulting in radioactive particles, such as Pi-mesons hitting the surface of the plant. The Earth passes through meteor showers, comet tails and asteroids. Objects hit the Earth that originated on other planets such as Mars and it is not fanciful to point out that our solar system will brush against the interstellar dust from other systems and even other galaxies.

Amongst this debris are definitely organic molecules, and it may even include organisms that have 'evolved' away from the Earth and outside our solar system.

In economic terms are we living in a closed sytem or do, in fact, new products that are completely unexpected suddenly appear? Think of pocket calculators and computers and the absolute plethora of new toys, new magazines, new books and music that have appeared in the last few years and were not predicted by anyone. Also consider how new markets for old products arise unexpectedly.

If this is the case and the Earth is not a closed system could it be that the level of complexity we have now is the greatest that it has ever been? Could there now exist more individuals and more species than ever before? Could the variety of products that are offered to us for sale be greater than previously? Could people generally be better off than they were?

It is clear that evolution and survival of the fittest is by no means the only method for development of species and it is also true that the market does not necessarily know best. In fact chaos, catastrophe and complexity theories must all be taken into account when we consider any complex situation, and we must cast aside our assumption that we represent the pinnacle of evolution and that our present economic system is the best that can possibly be achieved.

Perhaps greater successes are around the corner as we near the end of the millenium but we must try very hard to avoid catastrophe. Continuously blinding ourselves to facts will certainly not assist us.

Fred Hoyle and Chandra Wickramisinghe present their evidence in this book in an understandable and entirely readable form. It is up to you to judge the veracity of their ideas.

PROLOGUE

When sightings of unidentified flying objects were claimed in the years following the Second World War it seemed that genuine occurrences, like the reflection of sunlight from an escaped weather balloon, would soon be separated from the rest, the rest being nothing but a motley collection of inaccurate observation, hysterical reports and deliberately planned fakes.

This did not happen, however. In 1950 people wanted to believe in UFOs, but they didn't. Nowadays, people want to believe in UFOs, and they do. There is a widespread feeling that the time is ripe for something of the kind to be proved true, and in a certain sense we shall demonstrate in this book that it is indeed true.

However, the Earth is not being invaded by intelligent beings in spacecraft. Such a concept is but a crude perception of the real situation; it is to 'see through a glass darkly', just as the poet who is struck by the beauty of a rose or the majesty of the starry sky perceives only glimpses of far deeper wonders.

A rose carries within it a heritage that stretches across aeons of time, a heritage which connects its fleeting present existence to the most remote past, and to distant far-flung parts of the Milky Way – our galaxy. So it is for every speck of living matter on the Earth, including of course ourselves. In the starry vault above us is writ, not merely the story of life on our planet, but of life throughout the Universe.

The oxygen, carbon nitrogen, phosphorus, sulphur, and the metals (to name several of the more important kinds of atom of which living material is composed) had their origin in the stars. This much is well-known. In this book we shall go further by showing that the exceedingly complex arrangements of these atoms in living material also had their origin on a cosmic scale. We shall find that recent developments in astronomy and biology make the case for life as a galaxy-wide phenomenon quite overwhelming.

Let us here state the questions for which we seek the answers:
● How uniform is the basic chemical fabric of life?

- Did life start here on Earth spontaneously, or was it brought to the Earth from outside?
- What is the evidence for bacteria and other micro-organisms existing outside the Earth? And how do the recent NASA find ings about life on Mars tally with this evidence?
- Is life still arriving to the Earth from outside, and how does such life interact with lifeforms on Earth?
- How frequently might we expect life to turn up elsewhere in the Universe at large?

We ourselves have been addressing questions such as these from the early 1970's and have published a long series of articles in scientific journals as well as in books describing our progress. The present volume contains an update of our thoughts on the cosmic nature of life, and includes a selection of readings from our earlier publications that are not easily accessible at the present time.

We hope the reader will find that answers to the above questions, which have long been thought unanswerable, became almost self-evident as soon as the relevant facts are put together in an appropriate way. Despite a steady accumulation of facts over a hundred years that pointed to life being a cosmic phenomenon, orthodox opinion maintained stubbornly that life was of a purely terrestrial nature. The situation is strikingly reminiscent of the resistance in the 15th and 16th centuries to accept facts that related to a sun-centred planetary system. Man, having been forced to accept a trivial position in the Universe in regard space and time, appears to have clung tenaciously to the idea that life at least is Earth-centred. This situation is rapidly changing, however. There are on-going programmes to search for extraterrestrial intelligence, using radio telescopes, and serious claims have recently been made by NASA that a meteorite from Mars carries possible evidence of past life on that planet. A long overdue paradigm shift may well be in sight. If this happens its importance could be on a par with the discoveries of Kepler, Galileo and Newton in the 15th and 16th centuries, and those of Wallace and Darwin in the 19th century.

In a broader perspective, the acceptance of the idea of life originating outside the Earth, and existing outside the Earth would have profound implications for the future progress of Mankind.

From the exalted status of being the pinnacle of life on the sole inhabited planet in the Universe, Man would be forced to accept the inferior role of being just another life form on a quite ordinary planet, the Earth, which is one of many billions of similar planets orbiting countless stars in the Universe. Our perceptions of ourselves, our sense of self-importance and self-esteem would be dramatically altered. At long last we might be forced to accept our true ancestral links with the cosmos, hopefully leading to the emergence of an all-encompassing cosmic world view. Such a world view, in the fullness of time, might be expected to give rise to a new social order. The realisation that we are not alone as living creatures in the vast universe may help us achieve an inner peace, and more importantly, perhaps, guide us to live in peace with our terrestrial as well as our cosmic neighbours.

1

PLANETS AND LIFE

A tour through the Solar System

1.1. THE EARTH

Our geological history starts with the formation of the Earth's crust about 4.5 billion years ago. The Sun had by then shrunk to approximately its present size and had begun its long life as a main sequence star. It had just started converting hydrogen to helium in its interior and at this early stage of its life the luminous output was some twenty-five per cent less than it is now. As a consequence of the lower luminosity, the inner planets were colder than they are at present. The Earth was also dry and heavily cratered from the impacts which it had suffered during the final stages of its aggregation. In many respects our planet would have looked similar to the present-day Moon.

How did such an arid, dreary, lunar-like landscape become transformed into a habitat for life? At the time when the formation of the Earth was largely complete, the aggregation of the outer planets had barely started. The region between the present orbits of Uranus and Neptune was occupied in the main by a multitude of comparatively small objects, about a hundred kilometres in diameter, orbiting around the Sun. These objects, which we can identify nowadays as giant comets, were made up of a mixture of ices which condensed in the outflowing planetary gas, together with quantities of interstellar matter – dust and molecules –that became mopped up from the dense interstellar material in which the new solar system was embedded.

About half of these objects went to form the outer planets; the remainder was flung out partly into interstellar space and partly to form the reservoir of comets known as the Oort cloud. This cometary reservoir now surrounds the entire solar system at distances upto about a light year from the Sun. Before 4 billion years ago when the outer planets were being accreted through collisions between giant cometary objects, the inner planets would also have been subject to severe bombardment by fragments of these comets. Studies of radioactivity in moon rocks have shown that violent bombardment events did indeed continue on the lunar surface up to about 3.9 billion years ago. It is difficult to imagine that similar events did not occur at the same time on the Earth's surface, and it is now generally accepted that our planet was unfit for the widespread dispersal of life much before this time.

When the most violent phase of comet impacts had ceased, some 3850 million years ago, the geological record also reveals the earliest indirect evidence of terrestrial microbial life[1]. This evidence is in the form of a small fluctuation in the ratio of carbon isotopes in carbonate deposits. It is well-known that carbon has two stable isotopes ^{12}C and ^{13}C, with ^{12}C being about 90 times more abundant than ^{13}C on the Earth as a whole. Photosynthetic organisms, which convert carbon dioxide in the atmosphere into biological material, have a slight preference for $^{12}CO_2$ compared to $^{13}CO_2$. Thus biologically generated hydrocarbons and carbonates have a small deficit of ^{13}C compared with carbonates that are generated by non-biological means. Finding a deficit of ^{13}C relative to ^{12}C in ancient sediments, typically about twenty parts in a thousand, can therefore be taken as evidence that photosynthetic life was present when the sediments were laid down. This indeed has been found in the oldest terrestrial sediments post-dating the intitial bombardment, showing that photosynthetic microbial life already existed 3850 million years ago. Thus life appeared on Earth as soon as the conditions for life became prevalent with little or no delay.

The Earth's atmosphere evolved significantly over the half billion years that preceded the cessation of impacts, the atmosphere probably being derived from the materials supplied by the impacting cometary objects, and with water also being derived from the same source. However, the water would at first have been frozen,

[1]Mojzsis, S.J., et al, *Nature*, 384, 55, 1996.

but cumulative additions of carbon dioxide could have rapidly changed the terrestrial climate.

The total amount of carbon dioxide now present as carbonates in sedimentary rocks must have initially come from comets. There could well have been a carbon dioxide pressure of many atmospheres over the primitive Earth, a circumstance that would quickly have changed our abode from being cold and dry to becoming hot and humid. This is because carbon dioxide in the atmosphere produces a 'greenhouse' effect. Visible light from the Sun is able to pass through the atmosphere and to heat the Earth, but infrared radiation from the heated Earth is impeded from escaping because atmospheric carbon dioxide is able to absorb it. The effect is to lift very considerably the temperature at ground level. As the temperature rises, water-ice melts, water vapour evaporates into the atmosphere, and the water vapour pressure increases steeply with the temperature. Convection sets the atmosphere into motion with rapidly rising and falling columns, enabling water vapour to condense into droplets high up in the atmosphere. The effect of this condensation is to produce a cooling from the upward transfer of latent heat. Such a cooling process, however, would not have become critically important until the oceans had become very warm, perhaps with a temperature above fifty degrees centigrade. This would have favoured the more thermophilic of the cometary organisms, namely those which originated in water that was warmed considerably by the release of chemical energy. Such microbial types may be identified with populations of archaebacteria that have recently been found to inhabit thermal vents in the oceans.

The first direct evidence of microbial fossil structures in the rocks shows up some 200 million years after the occurrence of isotopic signatures, 3600 million years ago, and by this time a remarkable diversity of microbial types seems to have been achieved. The speed with which life became widespread and diverse on Earth implies that either the original genesis and subsequent evolution of life is a very likely and simple event when the conditions are right, or that life was seeded as a wide range of different types from already existent life in the cosmos. The former appears not to be the case since the creation of life from non-life has never been shown to be even partially possible in the laboratory. In Chapter 2 we shall show that it is vastly more likely that the second possibility is the case, and that life is widespread in the cosmos and that a wide range of different types of microbial life from the cosmos was injected to the Earth along with the cometary volatiles, and that such life simply took root and multiplied at the very first moment when violent impacts had stopped, and a stable terrestrial crust and atmosphere had been acquired..

The eventual cooling of the Earth would have depended critically on the rate at which atmospheric carbon dioxide became converted into limestone in rocks. Carbon dioxide can be 'fixed' into rocks by dissolving in water and by the water then coming in contact with suitable solid material. At the present time these processes are greatly helped by living organisms which can concentrate carbon dioxide in the soil. The carbon dioxide is taken up by rainwater, which in turn washes over rocks to produce deposits of limestone. These processes were not available at the beginning. Besides, carbon dioxide loses its solubility in hot water, and so for a combination of reasons the fixing of the carbon dioxide in the primitive atmosphere could only have been a slow process.

Slow as it may have been, the fixing of carbon dioxide eventually took place. As the pressure of atmospheric carbon dioxide

Table 1.1
Approximate temperature upper limits for different groups of organisms (after T. D. Brock)

Organism	Temperature upper limit ($^{\circ}$C)
Nonphotosynthetic prokaryotes (bacteria)[1]	<90
Photosynthetic prokaryotes (blue-green algae)	73-75
Eukaryotic microorganisms (certain fungi and the alga *Cyanidium caldarium*)	50-60
Animals, including protozoa	45-51

[1]Recent data has shown that some microorganisms can live at temperatures of 113°C (*Science*, 275, 933, 1997).

declined, so did the temperature at the Earth's surface. There is therefore an interesting and potentially informative relationship between early life forms and temperature, especially when we make comparisons with modern life forms, of which T. D. Brock has made valuable studies at the Yellowstone hot springs (Table 1.1).

There is a strong indication that the Earth did not cool below fifty degrees centigrade until about a billion years ago, when the first eukaryotic microorganisms became evident. More complex life forms may have evolved late on the Earth for the simple reason that the temperature was previously too high.

This picture would be consistent were it not for fragmentary evidence of glaciations occurring at much more remote epochs in the pre-Cambrian era. A tillite (glacial deposit) at Gowgandra in the Timiskaming subprovince of Ontano points to the occurrence of glaciation having occurred as long as 2.5 billion years ago. Of course one might argue that the Earth has received repeated additions of volatiles, that the first fixing of carbon dioxide had indeed occurred by two and a half billion years ago, and that a subsequent high temperature phase was caused by a later acquisition of carbon dioxide. That the present picture is basically correct, however, is shown by two circumstances. In the first place, there is the correlation we have already noticed between temperature and age for the earliest life forms (Table 1.1 and Fig. 1.1). Secondly, there is the low luminosity of the Sun. Without some strong counter-effect like a

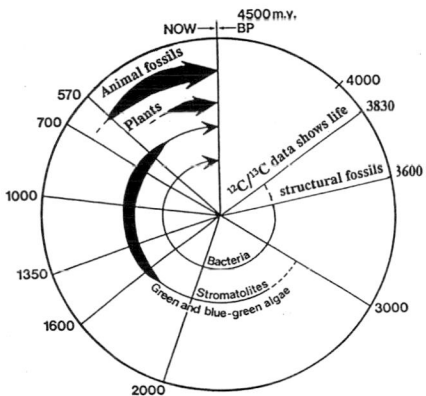

Fig. 1.1 Life forms in the geological record (BP = Before Present; m.y. = million years).

high pressure of carbon dioxide, there would have been a perpetual
worldwide glaciation (because of the low solar luminosity), which
certainly did not happen, since the oldest rocks with ages of about
3.8 billion years are known to have been in contact with liquid
water.

The best explanation therefore of the known facts relating to
the origin of life on the Earth is that in the early days material from
comets brought about the spreading of water and other volatiles
over the Earth's surface. Then about 4 billion years ago life also
arrived from a life-bearing comet. By that time conditions on the
Earth had become sufficiently similar to those on the cometary
home for life to be able to persist here, probably at first tentatively
and then with some assurance as time went on. The long evolution
of life on the Earth had begun.

1.2. HABITABILITY OF PLANETS AND SATELLITES
Could life exist on other planets?

The scientific exploration of the planets in the solar system has been
pursued with a variety of available techniques – optical, ultraviolet
and infrared measurements, radar and radio astronomy – and yet
some of the most startling discoveries have come and are still coming from direct explorations with spacecraft. In this section our
main concern will be to review the recent explorations of the solar
system, and in particular to examine the question of whether there
is evidence for life, or the potential for any form of life, on other
planets or on planetary satellites within the solar system.

Recent developments in microbiology have unravelled an
almost uncanny range of survival properties for bacteria.[1] Viable
microbes have been recovered from drills of the Earth's crust to
depths of 8 kilometres, from Antarctic ice, as well as from the
stratosphere. Whilst the number of culturable microbial species is
no more than a few thousand, the existence of millions of 'dormant'
species have been inferred from studies of bacterial DNA in a

[1]Postgate, J., *The Outer Reaches of Life* (Cambridge University Press, 1994).

variety of terrestrial samples. The limits of life appear to be a fast-receding horizon, giving confidence to search for extraterrestrial life even in locations that might at first sight appear hostile.

If, in our search for life, we move outwards from the Sun along the sequence of the planets, the first one we reach is Mercury, which is also the smallest. With a surface gravity about one third that of the Earth and with an orbital period around the Sun of eighty-eight Earth days, Mercury appears to an observer on the Earth to have phases similar to those of the Moon. From the surface of Mercury the Sun would look nearly three times larger than it does from the Earth, and the solar radiation received on a given area would be nearly ten times greater than that received by a similar area on the Earth. The period of rotation of Mercury about its axis is about fifty-nine Earth days and the surface temperature varies between about 400 degrees centigrade on the sunward side to about -200 degrees centigrade on the dark side. This intense daytime heat combined with the planet's low surface gravity is the reason why Mercury has no atmosphere beyond a tiny amount of the heavier rare gases. With a surface which probably resembles that of the Moon (on which of course there is no life), Mercury is perhaps the least hospitable of the inner planets, and there can be little hope of finding life there.

Venus

Moving further out from the Sun, we come next to Venus. Venus is very similar to Earth in size and mass, and when they are at their closest the two planets are only twenty-five million miles apart. Venus has an unusual brightness which is caused by the high reflectivity of the thick clouds shrouding its surface, and it has phases like the Moon. At maximum brightness it has a brilliance which is exceeded only by the Sun and Moon. The most accurate information concerning the atmosphere of Venus came from a series of space probes. The composition of the atmosphere is 97 per cent carbon dioxide, 1 or 2 per cent carbon monoxide and very little oxygen or water. The top of the thick cloud cover on the planet is about forty miles above the surface. The temperature at the surface is about

500 degrees centigrade and the pressure of the atmosphere at the planet's surface is about a hundred times that of the Earth at sea level. These conditions do not look very promising for life, at any rate on the surface.

What are we to make of these surface properties of Venus? Could we rule out the existence of life higher in the atmosphere where the temperatures and pressures might be more favourable? The high surface temperature of Venus is caused by the 'greenhouse effect' due to the thick overlying layer of carbon dioxide. Sunlight filters through the atmosphere and heats the surface, but the infrared radiation emitted from the hot surface is trapped because of the greatly broadened absorption bands of carbon dioxide. The trapping of infrared radiation lifts the surface temperature in much the same way that the primeval Earth may have been heated by a carbon dioxide atmosphere, before the carbon dioxide on the Earth became fixed into limestone. The difference in the case of Venus is that there is little or no water to drive a convective refrigeration system such as we described in the last section. The temperature distribution in the atmosphere of Venus therefore appears to preclude the possibility of life except possibly in highly localised domains underneath the Venusian clouds where a water cycle might conceivably exist.

Mars
Little green men or patches of lichen?

On the journey outwards from the Sun, we bypass the Earth and come to the last of the four terrestrial planets, Mars (Plates 1 and 2). For a long time it attracted more attention than any other planet because it was considered a likely habitat for intelligent life. With a radius of about half that of the Earth, and a mass of approximately one-ninth, Mars has a surface gravity which is a little less than half that of terrestrial gravity. The Martian day is almost exactly as long as an Earth day, and because the tilt of its axis of rotation is the same as that of the Earth, the seasons are also similar to terrestrial seasons. On the other hand, Mars is further than the Earth from the Sun, so that the Martian year is nearly twice as long as the terrestrial year.

Speculations about intelligent Martian life arose as a result of observations of many enigmatic features on the planet's surface. There are light and dark patches which show irregular as well as seasonal variations, and the dark markings were at one time widely believed to have a green colour which was attributed to vegetation. We know now, however, that the patches appearing on Mars are in fact red rather than green, and that they are caused by the redistribution of clouds of fine dust particles over the surface. There was also a long controversy over the alleged Martian 'canals'. The Italian astronomer Giovanni Schiaparelli (1835-1910) saw what he thought was an elaborate network of lines criss-crossing over the planet's surface. These so-called canals were assiduously mapped by several observers, the most notable among them being the American astronomer Percival Lowell (1855-1916).

Although evidence for intelligent Martians remained tenuous, it was at the same time difficult to disprove the theory for many years. An argument that could not be refuted was that if the Earth were viewed optically from a Martian vantage point, our planet would prove just as elusive over the presence or absence of intelligent life. There was no way of resolving this question unequivocally until the first Mariner probes sent back close-up pictures of the Martian surface. The answer was of course disappointingly negative. Not only was there no sign of intelligent life, but there were no structures that even vaguely resembled the fabled canals. In fact the Martian landscape looked incredibly barren and uninviting.

Mariner 9 provided the first detailed survey of the whole of the Martian surface photographically with a resolution of one kilometre, and a few per cent of the surface at a finer resolution of a hundred metres. At this latter resolution any artifacts of an intelligent civilization should have been clearly visible, but nothing of the kind was seen. Instead, it showed that the Martian surface was covered with moon-like craters and ridges. Martian craters bear a general similarity to lunar highland craters, except that they are much shallower, and they were almost certainly caused by meteoritic impacts. Some of the largest of them may be several billion years old. There is also a set of curious-looking hummocks which are a few kilometres across at the base and about a kilometre high. They are most likely to have formed by a natural physical process involving the erosion of mountains by high winds, the wind speeds on Mars being much higher than on Earth.

The Mariner probes have also shown evidence for volcanic activity at certain sites on Mars and they have given us information about the behaviour of dust storms which occur sporadically over large areas of the surface. These storms give rise to conspicuous changes in the planet's coloration as viewed from Earth. A storm which occurred in 1971 was particularly violent and widespread. In certain localized areas such as the 'dust bowl' Hellas there may be perpetual dust storms (see Plate 2).

The equatorial surface temperature of Mars varies between a daytime high, which is close to the melting-point of water-ice, and a night-time low of about -100 degrees centigrade. (At the two landing sites of Viking 1 and Viking 2 the temperatures ranged from a high of -31 degrees centigrade to a low of -84 degrees centigrade.) The Viking 2 orbiter recorded a north pole temperature of −68 degrees centigrade. At this temperature carbon dioxide would not be frozen, so that the frozen polar cap material which was seen is believed to be mostly water-ice.

With the possibility of an advanced life form excluded, what are the prospects for finding any form of life at all on Mars? What about humbler life forms such as microbes or algae? The physical conditions on Mars turn out to be highly restrictive, and so we cannot be over-optimistic. The Martian atmosphere is mainly comprised of carbon dioxide (about 95 per cent), argon (about 3 per cent), nitrogen (1.5 per cent) and oxygen (0.15 per cent), together with a small quantity of water. The atmospheric pressure at the surface of the planet is about half a per cent of that on Earth. Such a thin atmosphere provides no protection from the Sun's ultraviolet radiation, and the power of this flux reaching the Martian surface would prove lethal to most, if not all, terrestrial organisms. On Mars, therefore, life could exist only in niches where a natural shelter is provided from ultraviolet light. Dust basins are possible sites, since their bases may be quite well shielded by the efficient extinction properties of fine dust. It could be that organic molecules associated with pockets of microbial activity are swept up in such dust storms, as we shall note later, and that the overlying dust protects these molecules temporarily from destruction.

A major goal of the space probes, Viking 1 and Viking 2, which landed on Mars on 20 July and 3 September 1976, was to search for life. A photograph of the landing site is shown in Plate 1. The probes were equipped to carry out biological experiments *in situ* on samples of soil, some of which were taken from under surface rocks. The presumption was that any microorganism which may be present had metabolic processes broadly similar to those of terrestrial microorganisms. The soil was treated with various nutrients, and expelled gases and the soil itself were examined in several ways. The results of the experiments turned out to be somewhat confusing. The soil was much more *active* than any known terrestrial soil and it has been described by chemists as a super-oxidant. One curious fact is that the *bioactivity* of the Martian soil apparently persists even after the soil is heated to well above normal sterilization temperatures. Another remarkable fact is that the Marian soil did not show detectable amounts of even simple organic compounds. This would mean that Martian microbiology, if it exists, must be remarkably thermophilic (heat-loving), and Martian ecology would have to provide a highly efficient scavenging system for free organic molecules. The situation turned out to be complex beyond all expectation and the outcome was highly uncertain. But to err on the side of caution NASA scientists announced in 1976 that the Viking results were inconsistent with life, and that some other explanation was required to explain the vigorous gas release that was observed.

We ourselves never regarded this result as being conclusive. In 1977 H. Abadi and N.C. Wickramasinghe showed that the absorptive index of dust particles released in the 1971 Martian dust storm had an ultraviolet spectral signature at 2200A, that we noticed was characteristic of biomaterial. We wrote in *Nature* (Vol. 267, 687-688, 1977) that:

> Dust sucked up into the atmosphere from large areas of the planet surface would have been witnessed in the 1971 dust storm, representing

a much broader sample of Martian surface material......More accurate
and extensive spectroscopic studies of Martian dust clouds may well
hold vital clues which have a bearing on the question of complex
organic molecules and primitive life forms on Mars . . .

In 1986 a careful re-examination of the Viking data combined
with nearly a decade of laboratory experimentation led to the startling (but little publicised) conclusion that primitive life could well
exist in subsurface niches on this planet. Not only was it the case
that the Viking experiments were not tested beforehand on the
most inhospitable terrain on the Earth – the dry valleys of the
Antarctic – until well after 1976, but when they were so tested
results identical to the Martian results were obtained. The Antarctic
dry valleys surely contain populations of bacteria, but their turn-
over rate was so small that no organics were detectable by the
Viking instruments. Levin and Straat, two scientists on the Viking
biology team, further embarrassed the establishment of NASA by
announcing in 1986 that all attempts to mimic the Martian results
using non-biological models were unsuccessful. Even more contro-
versially Gilbert Levin claimed that a series of colour photographs
of a Martian terrain at the Viking lander site, taken at regular intervals through a Martian year, showed seasonal changes in the colours
of rocks, fully consistent with the growth of lichens! What is true
beyond any doubt is that the results of the 1976 Viking experiments remain consistent with the presence of microbial life on
Mars. In 1996 this fact is being admitted by NASA at long last, and
in a forthcoming mission to Mars, a decade away, plans are afoot to
take care of the contingency that microbial life might be brought
back from Mars – even perhaps microbes that may be pathogenic
to human life!

The latest chapter in the exploration of Mars was opened in
August 1996 with studies of a 1.9 kilogramme meteorite (ALH
84001) which is believed to have originated from Mars. ALH84001
is just one of a group of meteorites discovered in 1984 in Allan
Hills, Antarctica, which is thought to have been blasted off the
Martian surface due to an asteroid or comet impact some 15 million
years ago. This ejecta orbited the sun until 13,000 years ago when

it plunged into the Antarctic and remained buried in ice until it was discovered. The presumed Martian origin of these meteorites (also known as SNC meteorites) seems to have been confirmed by several independent criteria. One that is perhaps amongst the most cogent involves the extraction of gases trapped within the solid matrix which were found to resemble in relative abundances the gases that were discovered in the Martian atmosphere. Also the ratio of oxygen isotopes $^{17}O/^{18}O$ in the mineral component is said to match the value found on Mars so closely that there is no reason to doubt a Martian origin.

A team of NASA investigators led by David S. McKay (McKay, D.S. et al *Science*, 273, 924, 1996) have found that within the meteorite ALH 84001 there are sub-micron sized carbonate globules (lower panel of Plate 3) around which complex organic molecules are deposited. The hazy rims seen around the carbonate globules are thought to be some sort of an organic layer, similar to the protective 'biofilms' that modern colonies of terrestrial bacteria are known to produce. The molecules that have been discovered around these globules include polyaromatic hydrocarbons (PAH's for short), which are characteristic products of the degradation of bacteria. Moreover, strings of ovoid shaped structures, such as are displayed in the upper panel of Plate 3. have been considered to be suggestive of fossilised microorganisms. Amongst the ovoid structures are also minute single crystals of magnetite, similar to structures that are commonly laid down by terrestrial iron-oxidising bacteria. Another indication that may point to biology is an enrichment of the carbon isotope ratio $^{12}C/^{13}C$ that has been found in the carbonaceous material of the meteorite. (We have discussed this effect already in another context.) McKay and his colleagues admit that their proposed identification involves a process of multi-factorial assessment. The totality of the available evidence, considered all together, in their view points to a microbial origin, although each single piece of evidence may be capable of more conservative interpretation.

An initial worry was expressed in some circles concerning the sizes of the presumed bacterial fossils. In general, they have dimensions that are some 5-10 times smaller than would be appropriate for normal terrestrial bacteria. However, it soon became clear, after looking through the work of Professor Robert L. Folk[1,2] of the University of Texas in Austin, that bacteria of similar sizes do indeed exist on the Earth and comprise a large class that has come to be known by the term *nannobacteria*. Folk and his colleagues have provided ample evidence to substantiate their claim that bacteria with diameters in the range 0.05-0.2 micrometres are largely responsible for mineral precipitation on the Earth.

It would indeed have been surprising if a claim of such profound importance as this was not immediately subject to meticulous scrutiny by the scientific community at large. Less than a year after the original discovery was published there have been sceptics as well as supporters who have come to the fore. Those sceptics who have voiced the opinion that PAH-type molecules could be non-biological have little to back them except the fact that interstellar clouds as well as normal carbonaceous meteorites are replete with similar substances. For reasons we shall outline in subsequent chapters (Chapters 2 and 3) it would appear far more likely that interstellar PAH's, in the quantities they are found, and similar materials in carbonaceous meteorites have a biogenic rather than an abiotic origin.

Another criticism has concerned the temperature of formation of the carbonate globules (lower panel of Plate 3). Two geochemists from the University of Colorado claimed that there was some evidence that the temperature at which the carbonate globules formed was far in excess of the survival limit for bacteria. But more recently, Dr. J.L. Kirschvink and his colleagues based at Caltech, have presented stronger evidence that points in a contrary direction. The carbonate globules had most likely been formed at temperatures well within the range for bacterial survival, even replication, and, moreover, it appears that ALH84001 had never been heated to above 110 degrees Celsius since 4 billion years ago, long before any bacteria could have infiltrated the rock.

A further point of contention relates to the origin of magnetite

---

[1]Lynch, F.L. and Folk, R.L., *Paper presented to the Annual Meeting of the Geological Society of America*, 1996.
[2]Folk, R.L., *Natural Science*, Vol 1, Article 3, 1997
(Internet – http://naturalscience.com.ms/ns.home.html

particles within the meteorite. J.P. Bradley and his colleagues have reported that some of the magnetite crystals are in the shape of whiskers that possess crystal defects, and that these are consistent with condensation from a high-temperature vapour. However, Dr. McKay has responded that his own team has not seen either whisker-like shapes or defects in the magnetite crystals in ALH-84001. And in any case one cannot yet rule out the possibility that biogenically produced magnetite could contain a small fraction of defects.

The debate continues. And it may do so for many years, perhaps until the next expedition to Mars brings back to Earth more decisive samples of microbial fossils or even viable microbes.

1.3. THE GIANT PLANETS AND THEIR SATELLITES:
A surprising possibility

When one considers the extremely wide range of conditions that support microbial life on Earth, the prospects for life on the satellites of Jupiter and Saturn cannot look bad. The relevant conditions include those prevailing at great depths in the Earth's crust, in the Antarctic ice shelves and in high temperature vents in the ocean floor. Similar conditions must occur quite plentifully upon the satellites of the giant planets.

The Jovian satellites were first examined at close range by the Voyager spacecraft in 1979, and they have come under even closer scrutiny following the launch of the Galileo orbiter in December 1997. Io, with a diameter of 3640 kilometres, has been found to be a hotbed of volcanic activity, some of which could be triggered by the build up of high pressure pockets of gas generated due to bacterial activity. A mosaic of cracks over the the icy surface of Europa (which has a diameter of 3130 kilometres) has led to speculations about the existence of a sub-surface ocean that harbours microbial life (Plate 4: top). These cracks are thought to be caused by tidal interactions with the planet Jupiter. Close-up views of the cracks were obtained by cameras aboard the Galileo orbiter early in

1997. Organic pigments appear to outline the cracks, some of which have soft dark edges, whilst others are interspersed with dark dots (Plate 4: bottom). This could imply that water laden with bacteria oozed out of the cracks and refroze in relatively recent times. The new discoveries give further credence both to the presence of a subsurface Europan ocean, and to the possible existence of lifeforms within it.

The most recent Galileo discoveries relate to the largest of the Jovian satellites, Ganymede. At close quarters areas of Ganymede's icy surface show a network of ridges and cracks generally similar to Europa. It appears that ice tectonics and volcanism has operated on this satellite as well, possibly mediated by biology. At the time of going to press Galileo spectrometers have detected the presence of complex organic cyanides on Ganymede, from which a connection with life could be inferred.

The Saturnian satellites have yet to be explored in detail. Of these, the largest satellite, Titan, with a diameter of 5150 kilometres and an atmosphere that includes hydrocarbons, must offer as good a prospect for life as any of the Jovian satellites.

1.4. EXTRASOLAR PLANETS
Is there evidence of planets around other stars?

Before attempting to find out about planets outside our own solar system, we first need to know how many stars may indeed be expected to have planetary systems. In looking for an answer to this question we can be helped by our understanding of the processes which led to the formation of our own solar system, for the same arguments can be applied to other stars. Unfortunately, theories of star formation and solar system formation are probably not as secure as many contemporary astronomers would like to believe.

The starting point of any reasonable theory of solar system formation is a fragment of an interstellar cloud. The formation of planets is an almost inevitable consequence of the formation of a solar-type star, for although most of the cloud mass goes into the star, a small fraction eventually breaks away to form the planets. Planet formation probably arises from a process whereby the bulk

Plate 1: Viking landing site on Mars. Searches for microbial life at this site in 1976 proved inconclusive. *(Courtesy NASA)*.

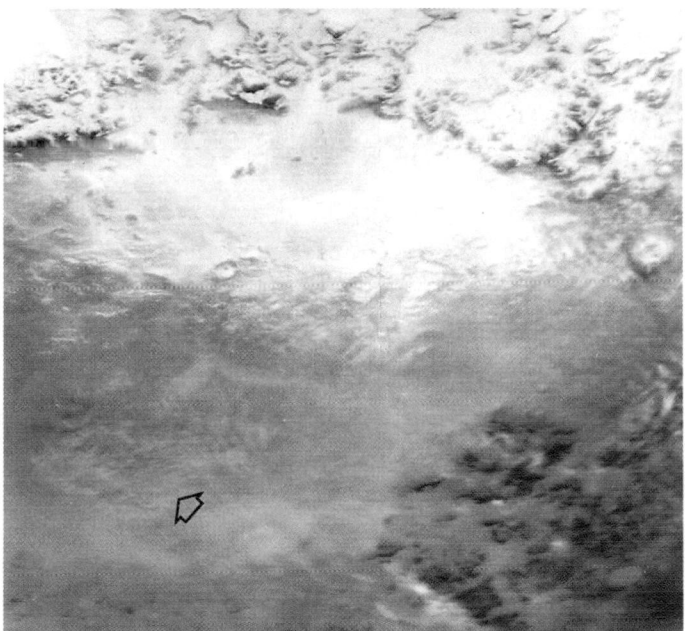

Plate 2: Dust storm in the Argyre Basin of Mars. *(Courtesy NASA)*.

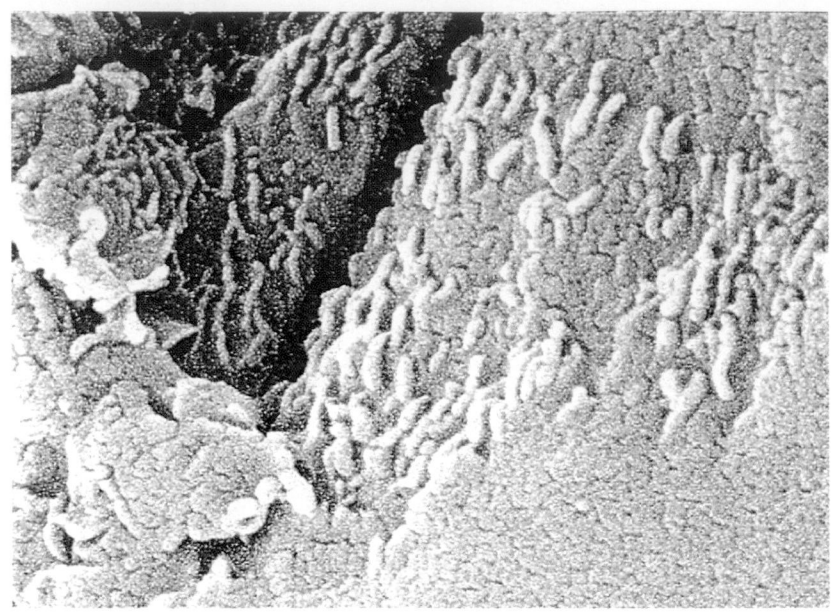

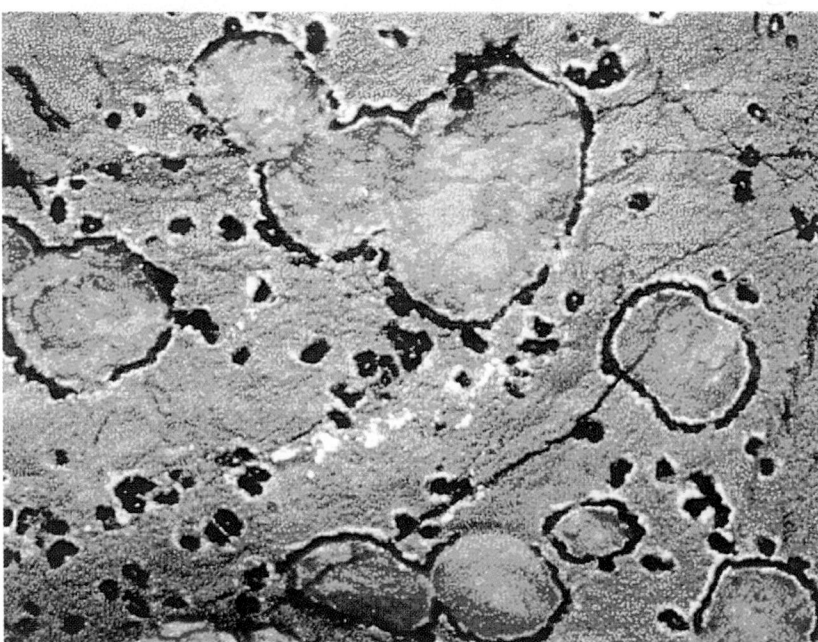

Plate 3: Evidence for microfossils in the Martian meteorite ALH84001. Upper frame shows elongated structures presumed to be fossilised nanobacteria. Lower frame shows larger carbonate globules around which complex organic compounds have been detected. *(Courtesy NASA)*.

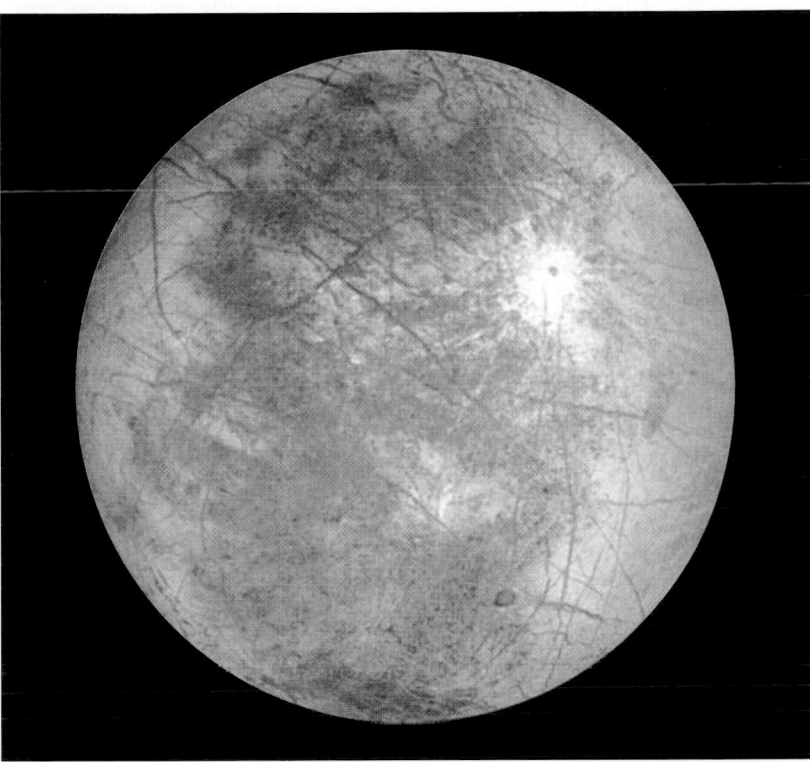

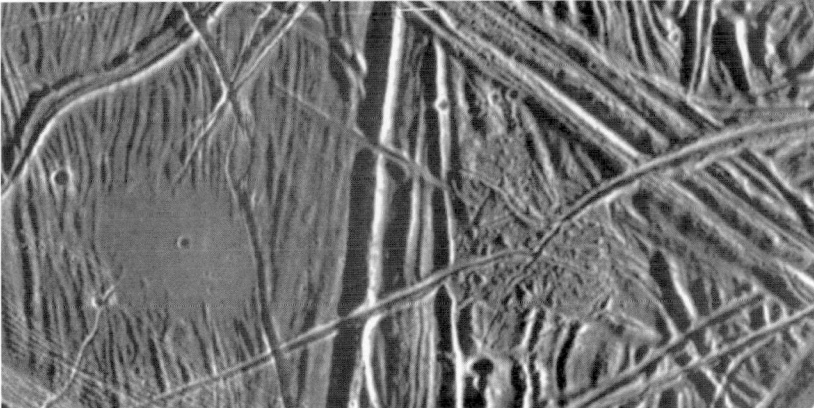

Plate 4: (Upper) Galileo image in true colours of the Jovian moon Europa, showing a criss-crossed network of fine cracks, coloured in an orange 'pigment'. Cracks are probably due to fractures in the icy surface that become filled with liquid from below, presumably laden with biological material. (Lower) A close-up of a portion of Europa's surface (18km wide) imaged on 20 December 1996 by the Galileo spacecraft. The sun lights the surface from the right and the resolution is about 26 metres. A smooth nearly rectangular area about 3km across, which is seen near the left of the picture, is probably caused by an episode of recent flooding from beneath the frozen crust. Notice that the pattern of ridges and grooves disappears over the smooth patch. Also compare the smoothness of this patch with the rugged nature of a patch of similar size further to the right. See also Plate 4a. (Courtesy JPL-NASA).

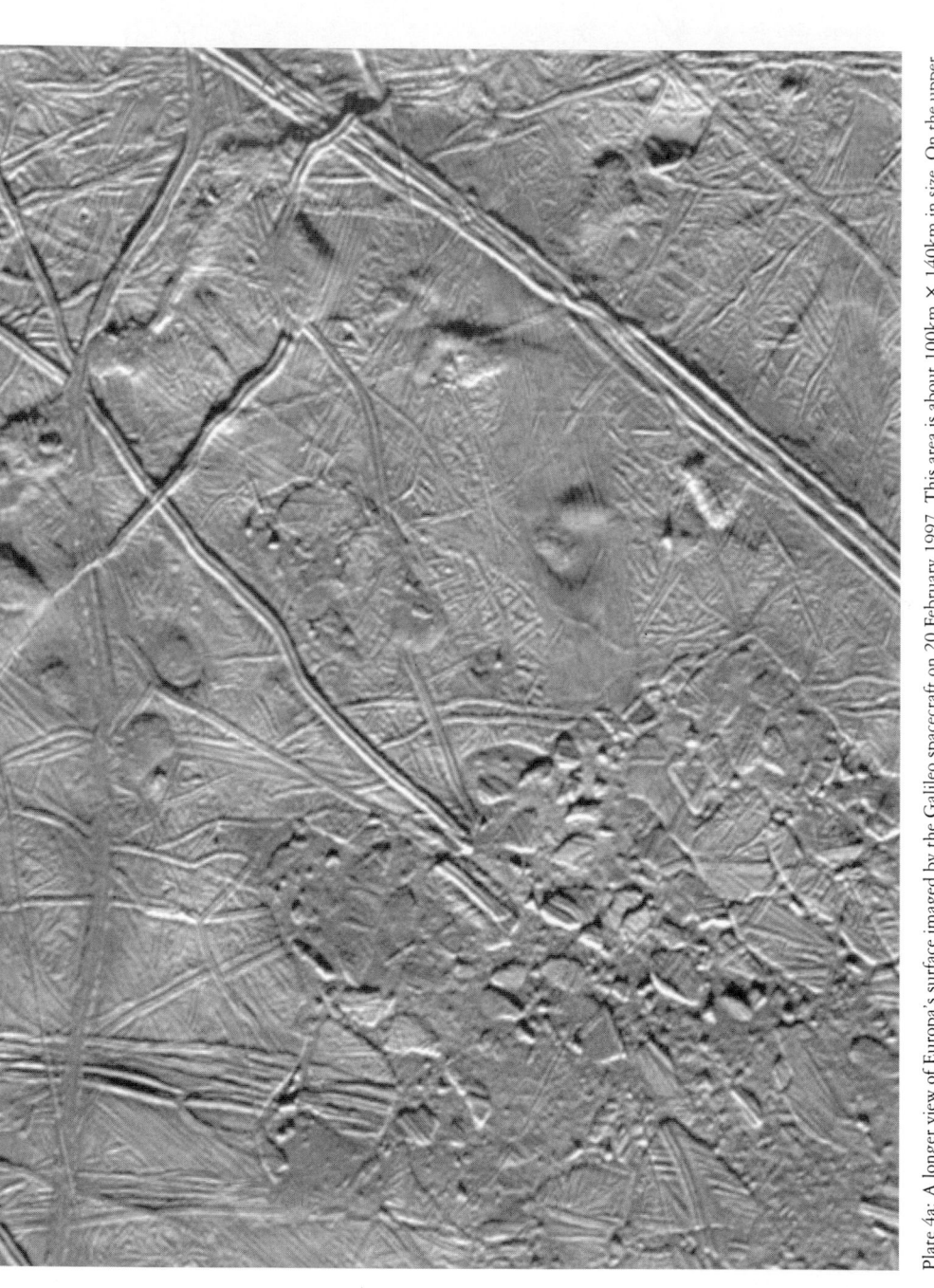

Plate 4a: A longer view of Europa's surface imaged by the Galileo spacecraft on 20 February 1997. This area is about 100km × 140km in size. On the upper right of this picture is an area that has been disturbed by an 'unknown process', resembling blocks of ice that float on the Antarctic during springtime. Note also the sprinkling of curious mounds enveloped by shallow depressions, probably representing sites of recent upwelling of liquid. (*Courtesy JPL-NASA*)

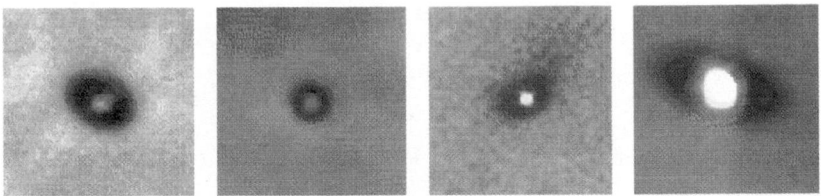

Plate 5: Hubble Space Telescope images of four protoplanetary discs around young stars in the Orion Nebula. *(Courtesy M. McCaughrean and C.R. O'Dell, NASA).*

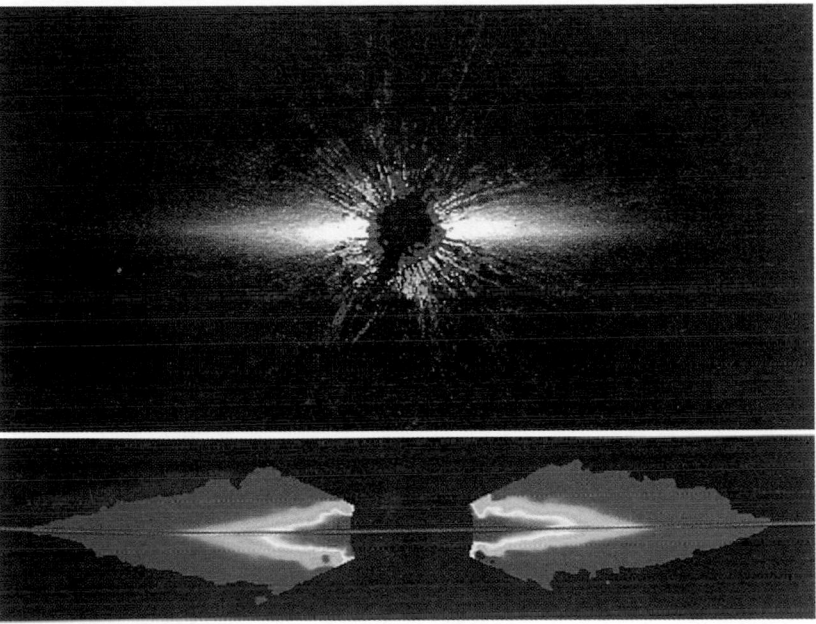

Plate 6: Hubble Space Telescope image of the inner regions of the dust disk around the star β Pictoris showing an innermost region free of dust. The formation of a planetary system is the most likely way to produce a central clearing such as this. The false colour image on the lower frame shows a warp in the dust disk towards the inner edge, thought to be due to the gravitational effect of planets. *(Courtesy C. Burrows and J. Krist, NASA).*

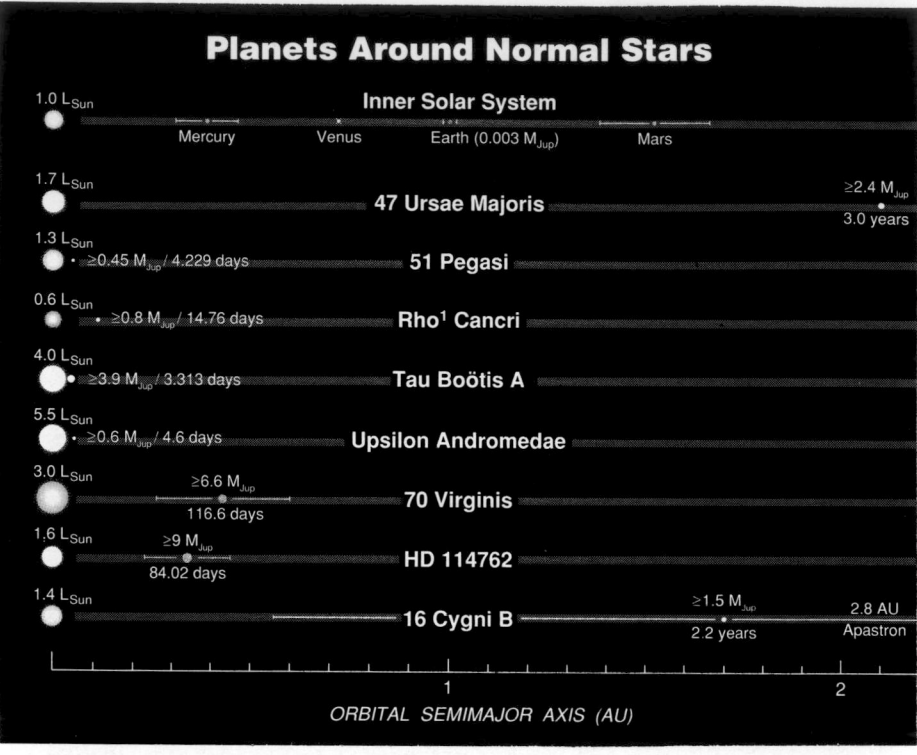

Plate 7: Summary of data on 9 recent detections of planets around nearby stars. The stars are scaled to their relative sizes. L_{Sun} is the luminosity of the Sun (Sol). M_{Jup} is the mass of Jupiter. Planets with highly eccentric orbits have horizontal bars showing maximum and minimum distances from the parent star. *(Courtesy G.W. Marcy, R.P. Butler and Sky and Telescope).*

Plate 8: The Orion nebula with giant dust clouds from which new stars have formed. (*Courtesy Anglo-Australian Telescope Board*).

Plate 9: A dense cloud of gas and dust in the nebula M16 from which new stars are being born. The stellar radiation is forcing its way out through the cloud at its edges. This Hubble Space Telescope image was taken by J. Hester and P. Scowen. *(Courtesy NASA)*.

of the star's angular momentum has to be shed along with a small fraction of its mass which forms the planetary disc. Four such discs are shown in Plate 5. This shedding of material has the effect of preventing a 'rotary crisis' which threatens to break up the star with too rapid a spin. This crisis does not overtake stars which are at least ten times more massive than the Sun, for these are better able to 'contain' their rotary forces. Stars of large mass are observed to spin rapidly and would not therefore be expected to have planets.

Our search for planetary systems should therefore be confined to stars less massive than about ten solar masses, though this still leaves the vast majority of stars as potential candidates. Another point about planets is that the central star, and hence the original cloud, must have basically the same chemical elements which are found in the Sun and which are essential for making up planets — oxygen, nitrogen, carbon, magnesium, silicon, iron and so on. As it happens, this condition is satisfied by the overwhelming majority of stars, so that planets with a composition like that of the Earth may well be common. If the genetic components of life are widespread as we shall be arguing in this book, potential sites for biological evolution could also be very numerous.

When it comes to the survival of life, long-term habitability and the emergence of a technological species, the conditions are much harder to fulfil. Evolution of even moderately complex life forms on Earth took over two billion years, yet a star which is fifty per cent more massive than the Sun will last only about the same time — two billion years — before swelling up to become a red giant star. This means that there would not be enough time for life to evolve on a planet around such a star, if our terrestrial experience is regarded as anything like the norm. In looking for habitable planets we should therefore perhaps restrict our attention to stars with masses less than about one and a half times the Sun.

Stars less massive than the Sun live longer on the main sequence, which means that the red giant phase does not intervene too early. Even so, planet-bearing stars of this type could have other restrictions and it is important to bear in mind the basic planetary requirements for habitability. In the first place, the planet must belong to a star which has a main sequence life of not much less than three billion years, and it must have an orbit that remains substantially unperturbed over this length of time. Secondly, the average temperature on the planet must lie 'comfortably' between the melting-point of ice and the boiling-point of water. This sets a distance range from the central star which in turn depends on the star's energy output. A third condition is that the planetary mass must be within a limited range (say from about half to two and a half times an Earth mass) so as to hold an atmosphere including water – in other words an atmosphere not very different from the Earth's. In the fourth place and finally, the planet's spin on its axis should be fast enough to minimize fluctuations of the temperature between day and night. A spin period which is much longer than several Earth days would normally be too slow, for it would produce unacceptable extremes of temperature.

What about the lower mass limit of a star, if its planets are to be habitable? The distance we need to be from a furnace in order to keep us warm depends on the intensity of the heat source – the more feeble the heat source, the nearer we have to be. In a very similar way, a life-bearing planet around a star of low mass, and hence low luminosity, would have to be closer to the star in order to maintain its average surface temperature in the 'comfortable' range. For stars less massive than seventy-five per cent of a solar mass, the star-planet distance becomes small enough for the effects of tidal friction to play a dominant role. This tidal interaction has the effect of greatly slowing down the rotation of the planet on its axis, so that the resulting extremes of temperature become unacceptable for the evolution of advanced life forms. This seems to whittle down the permitted stellar mass range to between three-quarters and one and a half times that of the Sun.

We have already seen that if a planet is to be habitable, its orbit must be substantially unperturbed for at least three billion years. This means, for instance, that we could not tolerate the Earth being slowly perturbed between the orbital radii of Mercury and Jupiter. This condition would not generally be fulfilled for stars which form binary or multiple stellar systems, because these stars pursue relative motions which would tend to perturb the planetary orbits periodically. These perturbations would make conditions on planets unstable and therefore inhospitable for life.

Of the two hundred billion or so stars in our galaxy, about eighty per cent fail to meet the conditions discussed above as being necessary for life. The remaining twenty per cent are not in multiple star systems and have masses in the appropriate range, three-quarters to one and a half times the mass of the Sun. The grand total of planetary systems in the galaxy capable of supporting life is therefore close to forty billion.

With so many possible planetary systems, should we not expect inhabited planets to be moving around some of the nearby stars? We certainly should, but before expanding on this question let us take an imaginary trip to our nearest stellar neighbour, Alpha Centauri, which is about four light years away. If we were then to look back on our solar system from the vantage point of a hypo-thetical Earth-like planet moving around this star, the Sun would be one of the brightest stars in the sky. If Jupiter, the largest planet, were on its own it would probably be just within the margin of detectability of a 200-inch telescope such as the one on Mount Palomar in California. As it happens, from the viewpoint of Alpha Centauri, Jupiter lies only about four seconds of arc away from the Sun's point-like disc, and the solar brilliance would make detection impossible if our present-day techniques were used. In a similar way reflected light from parent stars has so far made it impossible to directly observe planets outside our own solar system, though such observations could be feasible in the near future with the Hubble Space Telescope.

Another way of detecting planets moving around a nearby star is to look for dynamical effects on the visible star produced by invis-ible planets. If the Sun were observed from our hypothetic vantage point near Alpha Centauri, its apparent path in the sky would show a small regular wobble with a period of about twelve years, due to a slight pull by our largest planet Jupiter. The Sun and Jupiter move around their common centre of gravity with the orbital period of 11.8 years, and this motion produces an oscillation in the Sun's position in relation to the background of more distant stars. This oscillation might conceivably be observed from Alpha Centauri.

Astronomers have been searching for effects of this type in a number of nearby sun-like stars, and were beginning to get thoroughly frustrated when in 1995 Michel Mayor and Didier Queloz discovered a planet around the star 51 Peg in the constellation of Pegasus. (See review by S.V.W Beckwith & A.I.Sargent, *Nature*, **383**, 139, 1996). This was the first apparently secure detection of an extrasolar planet. The planet was estimated to have a mass of about half that of Jupiter but it is located only 0.05 astronomical units from the star, compared to Jupiter that is at some 5 astronomical units from the Sun[1]. Its orbital period is only 4 Earth days compared to 12 years in the case of Jupiter. Clearly this first extrasolar planet was far too hot for any life to survive at its surface. Since 1995 the race for finding planets has been gathering momentum. Shortly afterward Geoff Marcy and Paul Butler reported the discovery of two new planets, one orbiting 70 Virginis and the other 47 Ursae Majoris, lying respectively at 0.5 and 2.1 astronomical units from the parent stars. Next came detections of planets around ρ Canceri and τ Bootis, and the inference of a planet or planetary system within the dusty disc of β Pictoris from the Hubble Telescope images. (See Plate 6). The data relating to planets discovered to date are summarised in Plate 7.

   None of the planets discovered in 1995/96 around sun-like stars appears ideal for the maintenance of life, at any rate at their surfaces. At the present time the probability of finding a planet within the appropriate distance range for a 'comfortable' temperature, and with an appropriate mass range to retain a suitable atmosphere, may be estimated at about five per cent. This means that out of the forty billion possible planetary systems, only one in twenty may be congenial to life, suggesting as a very rough estimate that there may be an ultimate grand total of about two billion habitable planets in our galaxy. With a billion or so galaxies similar to our own in the observable universe, there would still be a staggering billion billion or so habitable planets in the observable universe.

   The average distance between stars in our immediate vicinity is about five light years. If as few as one per cent of these are attended by habitable planets, the average distance between such planets would be about twenty-five light years.

[1]One astronomical unit = average distance of the Earth from the Sun, about 93 million miles.

1.5. SEARCH FOR EXTRATERRESTRIAL LIFE

In attempting to determine the feasibility of communication with extraterrestrial life forms, it is necessary to estimate the average duration of a technology, though our experience in this context is of necessity limited. In all probability our own technological development is in its infancy, for although the evolution of life on Earth has gone on for at least three billion years, our technology – one that is only just capable of contemplating interstellar communication – is well under a hundred years old. If we believe the most pessimistic prophets of doom we cannot expect our technology to last more than three hundred years. The number of technologies at any time is given by the formula:

Number of technologies = Number of habitable planets ×
(Technological life-span/Main-sequence life of star)

With two billion habitable planets and an average main-sequence life of a star of ten billion years, this gives:

Number of technologies = Technological life-span in years ÷5

If three hundred years is taken as the typical technological life-span, we obtain only sixty technologies throughout the galaxy at any time. A much more optimistic estimate for our own technological duration is 300,000 years, which is about the length of time *homo sapiens* has been in existence. In this case the number of technologies in the galaxy turns out to be 60,000, with a grand total of sixty million million for the entire visible universe. Within the galaxy the average separation between technologies would be about two hundred light years. A direct two-way radio communication between neighbouring technologies would then take four hundred years. Our own 'intelligently' controlled radio transmissions have been underway for just 100 years and could therefore have reached civilisations within an expanding sphere presently measuring some 200 light years in diameter – just about half the average separation between technologies as calculated above.

However, we should note that technologies may or may not use radio waves for communication. One might imagine that higly advanced civilisations could use more targetted methods of remote communication such as laser beams.

In conclusion we note that there are serious difficulties in estimating the number of planets inhabited by intelligent beings in our own galaxy, let alone throughout the observable universe. In particular, we do not know how long our own technological civilization is going to last, or how typical it is. Even so, we have seen that statistically there must be very many habitable planets in our own galaxy. By looking more closely at certain aspects of the evolution of life we may be able to find out more about possible intelligent beings on other planets and to assess the practicalities of both space travel and communication.

2
PANSPERMIA: AN OVERVIEW

2.1. INTRODUCTION

A likely paradigm shift from a warm little terrestrial pond to an extraterrestrial source of microbial life has been signalled recently by NASA. When the evidence appeared in a scientific journal it was, however, disappointingly indecisive. NASA's campaign in the media can certainly be awarded high marks for public-relations skills. Linking the shift in a roundabout way to Mars can be seen to have been politically astute, because by tradition the public is attuned to the concept of life on Mars, attuned at any rate from the time of H.G. Wells. The point being that the world was becoming over-ripe for the shift and this seemed the smoothest way of bringing it about.

When paradigm shifts occur it becomes advantageous to those who were wrong before to pretend that the past did not exist, a process currently taking place in *Nature*, a publication with an unhappy record on this subject with a history extending back beyond modern times to the first editions of that magazine in 1869-70.

2.2. A SHORT HISTORY OF PANSPERMIA

The name panspermia was in our opinion ill-chosen and has probably done more to turn people off the concepts to which the name is currently applied than anything else. A better name is urgently needed. Even the crude bugs-from-space appellation is better (BFS). We might also suggest the term *cosmicrobia* for consideration, a word that combines both cosmic and microbial meanings.

Until the late 19th century panspermia meant the passage of organisms through the Earth's own atmosphere, not an incidence from outside the Earth. In this form it seems to have been used first by the Abbé Lazzaro Spallanzoni (1729-99). But almost a century before that, Francesco Redi had carried out what can be seen as a classic experiment in the subject. He had shown that maggots appear in decaying meat only when the meat is exposed to air, inferring that whatever it was that gave rise to the maggots must have travelled to the meat through the air.

A very long wait until the 1860's then ensued, until Louis Pasteur showed by experiments on the souring of milk and the fermentation of wine that similar results occurred when the agency passing through the air were bacteria, replicating as bacteria, not producing a visible organism like maggots. The world then permitted Pasteur to get away with a huge generalisation, and honoured him greatly both at the time and in history for it. Because by then the world was anxious to be done with the old Aristotelian concept of life emerging from the mixing of warm earth and morning dew. The same old concept was to arise again in the mid-twenties of the present century, however, but with a different name. Instead of Aristotle's warm earth and morning dew it became a warm organic soup. Thus proving that to the human mind naming things matters more than rational thought, a truth that is becoming an essential weapon in the armoury of every well-equipped scientist.

Pasteur's far-ranging generalisation was the precept that each generation of every plant or animal is preceded by a generation of the same plant or animal. This view was taken up enthusiastic-ally by others, particularly by physicists among whom John Tyndall lectured frequently on the London scene, as for instance in a Friday evening discourse at the Royal Institution on 21 January 1870. It was to this lecture that the editorial columns of the newly established *Nature* objected with some passion. Behind the objection was the realisation that were Pasteur's paradigm taken to be strictly true, the origin of life would need to be external to the Earth. For if life had no spontaneous origin, it would be possible to follow any animal generation by generation back to a time before the Earth existed, the origin being therefore required outside the Earth.

This was put in remarkably clear terms in 1874 by the German physicist Hermann von Helmholtz (in *Handbuch der Theoretische Physik,* Vol 1, transl. H. von Helmholtz and G. Wertheim, Braunschweig, 1874):

'It appears to me to be a fully correct scientific procedure, if all our attempts fail to cause the production of organisms from non-living matter, to raise the question whether life has ever arisen, whether it is not just as old as matter itself, and whether seeds have not been carried from one planet to another and have developed everywhere where they have fallen on fertile soil . . . .'

Sir William Thomson (Lord Kelvin) said of Pasteur's paradigm: 'Dead matter cannot become living without coming under the influence of matter previously alive. This seems to me as sure a teaching of science as the law of gravitation . . .'

So if life had preceded the Earth, how had it arrived here and where had it come from? Earlier in the 19th century the German physician R.E. Richter had suggested that living cells might travel from planet to planet inside meteorites. Inadequacies in Richter's dynamics permitted J. Zollner in the 1870's to raise objections, eagerly seized on by orthodox opinion. But Kelvin's superior knowledge of dynamics allowed him to see that there was nothing to Zollner's objections, in particular that evaporation from the outside of a large meteorite keeps its interior cool, thereby reasserting the possibility of organisms being carried from planet to planet inside meteorites. In his presidential address to the 1881 meeting of the British Association, Kelvin drew the following remarkable picture:

'When two great masses come into collision in space, it is certain that a large part of each is melted, but it seems also quite certain that in many cases a large quantity of debris must be shot forth in all directions, much of which may have experienced no greater violence than individual pieces of rock experience in a landslip or in blasting by gunpowder. Should the time when this earth comes into collision with another body, comparable in dimensions to itself, be when it is still clothed as at present with vegetation, many great and small fragments carrying seeds of living plants and animals would undoubtedly be scattered through space. Hence, and because we all confidently believe that there are at present, and have been from time immemorial, many worlds of life besides our own, we must regard it as probable in the highest degree that there are countless seed-bearing meteoric stones moving about through space. If at the present instant no life existed upon the earth, one such stone falling upon it might, by what we blindly call natural causes, lead to its becoming covered with vegetation.'

Thus almost 120 years ago the ideas recently put forward by a number of commentators in the wake of NASA's activities were already well-known. Demonstrating yet again the oft-repeated lesson that before advancing new theories it is as well to read what has gone before.

It is a weakness of science that unless an idea has a means of advancing itself through observation or experiment it stultifies, almost regardless of how good the idea may be in itself. Unfortunately there was no way at that date, 1881, whereby observation or experiment could be brought seriously to bear on Kelvin's formulation of panspermia.

The next facet in the story is associated with Svante Arrhenius, whose book *Worlds in the Making* appeared in English, published by Harpers of London in 1908. (A short paper dealing with this same topic was published in 1903.) Arrhenius' contribution rested on two main points, one good, one not so good. The good point was that microorganisms possess unearthly properties, properties which cannot be explained by natural selection against a terrestrial environment. The example for which he was responsible was the taking of seeds down to temperatures close to zero Kelvin, and of then demonstrating their viability when reheated with sufficient care. Many other unworldly properties have been discovered over the years, as for instance the ability of microorganisms to survive inside a nuclear reactor. With all these properties being highly relevant, however, to survival in space.

The not-so-good point was that Arrhenius conceived of microorganisms travelling individually through the galaxy from star system to star system. He noticed that organisms with critical dimensions of 1 micron or less are so related in their sizes to the typical radiation wavelengths from dwarf stars in the galaxy that radiation pressure on them can be strong enough to challenge galactic gravitation, permitting the particles to be scattered around here, there, and everywhere throughout the interstellar spaces. But individual particles can be attacked by interstellar ultraviolet light, which was already known in the first decades of the century to be capable of deactivating bacteria and other microorganisms.

An attack on Arrhenius' views was mounted in 1924 by P.

Becquerel (*Bull.Soc.Astron*, 38, 393, 1924), on the basis of ultraviolet damage and this attack was widely accepted and has been repeated many times since. This despite the argument being of poor quality, a sure sign of the correctness of the remark due to Julius Caesar: 'Men readily believe what they want to believe.'

Five important points can be put forward to support Arrhenius' views:

● A coating of only a few microns thickness of a carbonaceous deposit at the surface of a culture of a particular microorganism provides essentially total shielding against ultraviolet light.

● Microorganisms are not really killed by ultraviolet light, they are only deactivated. This happens through a shifting of certain chemical bonds contained in the genetic structures of the organisms. Without destroying the genetic structures themselves, permitting the original properties of an organism to be recovered, as it can be in many cases, once the source of ultraviolet radiation has been removed, through the operation of suitable enzymes.

● Some bacteria and other microorganisms possess an astonishingly unearthly resistance to ultraviolet light, not known and not suspected in 1924 at the time of P. Becquerel.

● Microorganisms normally thought to be highly sensitive to ultraviolet light can through repeated exposures to it be made just as insensitive as the resistant kinds -yet another unearthly property.

● Experiments for bacteria within a nuclear reactor have demonstrated enzymic repair against actual DNA damage in cases where it is estimated that the DNA experienced as many as a million breaks in its helical structure, the breaks undergoing enzymic repair back to a viable condition.

These are properties which Arrhenius did not know and which obviously support his position. Nevertheless, bacteria and other microorganisms exposed remorselessly to cosmic rays and to the

Table 2.1

Interstellar Molecules

Chemical symbol	Name of molecule	Year of discovery	Part of spectrum	Chemical symbol	Name of molecule	Year of discovery	Part of spectrum
Two atoms				C_3H	Tricarbonate monosulfide	1986	Radio
CH	Methylidyne	1937	Visible	HCCN	(Unnamed)	1991	Radio
CN	Cyanogen radical	1940	Visible				
CH^+	Methyladyne ion	1941	Visible	*Five atoms*			
OH	Hydroxyl radical	1963	Radio	HCOOH	Formic acid	1970	Radio
CO	Carbon monoxide	1970	Radio	HC_3N	Cyanacetylene	1970	Radio
H_2	Molecular hydrogen	1970	Ultraviolet	CH_2NH	Methanimine	1972	Radio
CS	Carbon monosulfide	1971	Radio	NH_2CN	Cyanamide	1975	Radio
SiO	Silicon monoxide	1971	Radio	H_2CCO	Ketene	1976	Radio
SO	Sulfur monoxide	1973	Radio	C_4H	Butadiynyl radical	1978	Radio
NS	Nitrogen sulfide radical	1975	Radio	CH_4	Methane	1978	Radio
SiS	Silicon sulfide	1975	Radio	SiH_4	Silane	1984	Infrared
C_2	Diatomic carbon	1997	Infrared	C_3H_2	Cyclopropenylidene	1985	Radio
NO	Nitric oxide	1978	Radio	CH_2CN	Cyanomethyl radical	1987	Radio
HCl	Hydrogen chloride	1984	Infrared	C_4Si	(Unnamed)	1989	Radio
PN	Phosphorus nitride	1987	Radio	H_2C_3	Propadienylidene	1990	Radio
NaCl	Sodium chloride	1987	Radio				
AlCl	Aluminium chloride	1987	Radio	*Six atoms*			
KCl	Potassium chloride	1987	Radio	CH3OH	Methyl alcohol	1970	Radio
AlF	Aluminium fluoride	1987	Radio	CH_3CN	Methyl cyanide	1971	Radio
SiC	Silicon carbide	1989	Radio	NH_2CHO	Formamide	1971	Radio
CP	Phosphorus carbide	1989	Radio	CH_3SH	Methyl mercaptan	1979	Radio
SiN	Silicon nitride	1990	Radio	C_2H_4	Ethylene	1980	Infrared
NH	Nitrogen hydride	1991	Ultraviolet	C_3H	Pentynylidyne radical	1986	Radio
				CH_3NC	Methyl isocyanide	1987	Radio
Three atoms				HCCCHO	Propynal	1989	Radio
H_2O	Water	1968	Radio	H_2C_2	Butatrienylidene	1990	Radio
HCO^+	Formyl ion	1970	Radio				
HCN	Hydrogen cyanide	1970	Radio	*Seven atoms*			
HNC	Hydrogen isocyanide	1971	Radio	CH_3C_2H	Methylacetylene	1971	Radio
OCS	Carbonyl sulfide	1971	Radio	CH_3CHO	Acetaaldehyde	1971	Radio
HS	Hydrogen sulfide	1972	Radio	CH_3NH_2	Methylamine	1974	Radio
C_2H	Ethynyl radical	1974	Radio	CH_2CHCN	Vinyl Cyanide	1975	Radio
N2H	Diazenylium	1974	Radio	HC_3N	Cyanodiacetylene	1976	Radio
SO_2	Sulfur dioxide	1975	Radio	C_4H	Hexatriynyl radical	1986	Radio
HCO	Formyl radical	1976	Radio				
HNO	Nitroxyl radical	1977	Radio	*Eight atoms*			
HCS^+	Thioformylium	1980	Radio	CH_3OHCO	Methal formate	1975	Radio
SiC_2	Silicon dicarbide	1984	Radio	CH_3C_3N	Methyl cyanoacetylene	1983	Radio
H_2D^+	(Unnamed)	1985	Infrared				
C_2S	(Unnamed)	1986	Radio	*Nine atoms*			
SiH_2	Silylene (unconfirmed)	1990	Radio	CH_3CH_2OH	Ethyl alcohol (ethanol)	1974	Radio
C_2O	Dicarbon monoxide	1991	Radio	$(CH_1)_1O$	Dimethyl ether	1974	Radio
				CH_3CH_2CN	Ethyl cyanide	1977	Radio
Four atoms				HC_2N	Cyanoetriacetylene	1977	Radio
NH_3	Amonia	1968	Radio	CH_3C_4H	Methyl diacetylene	1984	Radio
H_2CO	Formaldehyde	1969	Radio				
HNCO	Isocyanic acid	1971	Radio	*Ten atoms*			
H_2CS	Thioformaldehyde	1971	Radio	$(CH_3)_2CO$	Acetone (unconfirmed	1987	Radio
C_2H_2	Acetylene	1976	Infrared				
C_3N	Cyanoethynyl radical	1976	Radio	*Eleven atoms*			
HNCS	Isothiocyanic acid	1979	Radio	HC_9N	Cyano-octatetrayne	1977	Radio
$HOCO^+$	Protonated carbon dioxide	1980	Radio				
C_3H	Propynylidyne	1984	Radio	*Thirteen atoms*			
C_3O	Tricarbon monoxide	1984	Radio	$HC_{11}N$	Cyanotetracetylene	1981	Radio
$HCNH^+$	Protonated HCN	1984	Radio				
H_3O^+	Protonated water	1986	Radio				

background of starlight in open (unshielded) interstellar space must be subject to eventual destruction. Provided there is a source for them, microorganisms expelled from a suitable source into interstellar space will ultimately be deactivated along the lines just discussed. Then the deactivated particles will be subject to steadily increasing degradation, ending we think in a production of polymeric molecules, with a resulting situation of the kind detected by astronomers over the years from the late 1960's to the present. Table 2.1, shows the current score of known interstellar molecules, to which should be added polyaromatic hydrocarbons, the amino acid glycine and the molecule vinegar.

2.3. THE DEVELOPMENT OF OUR INTEREST IN PANSPERMIA

We ourselves did not approach panspermia from the biological point of view described above, but from an attempt to understand the nature of the interstellar dust grains that populate the Milky Way. These show up as a cosmic fog, dense enough in many directions to blot out the light of distant stars. The grains are much the same in all directions as we look outwards from the Earth. They are of a size that would be typical for bacteria, a fact we knew from the beginning of our investigations in the early 1960's. Initially we attached no importance to it, however, not at the time having interests in panspermia.

But a crucially important fact is that the amount of the grains is as large as it can be if essentially all the available carbon, nitrogen and oxygen in the interstellar material is in the grains. The amount is about three times too large for the grains to be mainly composed of the next commonest elements, magnesium and silicon, although magnesium and silicon could be present in the material of the particles, as would hydrogen, as well as many less common elements in comparatively trace quantities.

A number of inorganic molecules composed of H,C,N,O immediately suggest themselves, condensible into solids at temperatures typically of about 50 Kelvin, the usual temperature of the grains.

As for instance CH_4, CO_2, NH_3, N_2, C . . . During the decade from the early 1960's we examined the properties of substances such as these, comparing their electromagnetic properties against the formidable number of facts concerning the grains that were then being discovered from astronomical observations. The result was no really good correspondence to the data, only a rather mediocre fit. The correspondences between the assemblies of particles investigated and the observations could characteristically be lifted to a certain moderate level of precision but never, no matter how much effort we expended, beyond that level.

It was a turning point when we at last realised that there are other very different kinds of material which can be made from these same four commonest elements – C, N, O, H, namely organic materials. Of which there are a vast number of possibilities, making for a great number of further investigations that could be made. By the mid-1970's, the observations were covering a large range in wavelength, from 30 microns in the infrared through the near infrared into the visible spectrum and thence through the ultraviolet. Thereby providing a large number of constraints which a satisfactory theory of the nature of the grains had by now to satisfy.

The observations in the visible spectrum continued to cause trouble, as they had done all along, until one day it was decided to calculate the properties of particles that were hollow. Particles that were about 70 percent hollow gave very good results. This is what bacteria become when they are fully dried out. So now, a decade and a half after starting on the problem, we decided to try the view that the interstellar particles represent a graveyard of bacteria. The big attraction of the test was that it could be carried out in a very direct way. We did not have to draw up a catalogue of assumptions in making our calculations. We did not need for example to *assume* a size distribution for our supposed bacterial particles. We could use a known bacterial size distribution, making our investigation without assumption on this and a number of other issues. The result is shown in Figure 2.1, an explanation of which is given in the accompanying legend. Here at last is a good correspondence of the observational points to the calculated expectation of the model. In

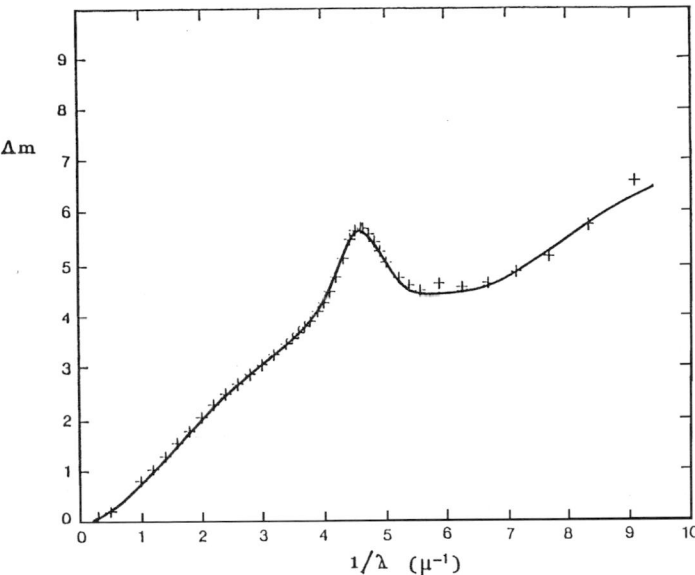

Fig. 2.1 Interstellar extinction curve (crosses) compared with bacterial model which combines the contribution from hollow bacterial grains and their degradation products, including aromatic molecules and viral sized fragments.

Fig. 2.2 is shown the expectation in the middle infrared spectral region, shown by the line again to be compared with the observational points. The observations are by D.T. Wickramasinghe and D.A. Allen; the bacterial data was obtained by S. Al'Mufti.

We were told by a number of chemists with experience of infrared spectroscopy that a curve like that of Figure 2.2 could be obtained from non-biologically derived organic materials in many ways. Since by now we had examined without success literally hundreds of infrared spectra of organic compounds, we did not believe this claim. Consequently, we asked the chemists in question that an explicit example be produced. But it never was, with the exception perhaps that some expensive laboratory experiments, involving carefully controlled irradiation of inorganic mixtures, have been claimed to yield undefined 'organic residues' that may possess some of the desired properties.

We all in some degree tend to think when we run into an apparently absurd proposal, that any form of opposition to it will do. Because in the end what is absurd will be proved to be absurd,

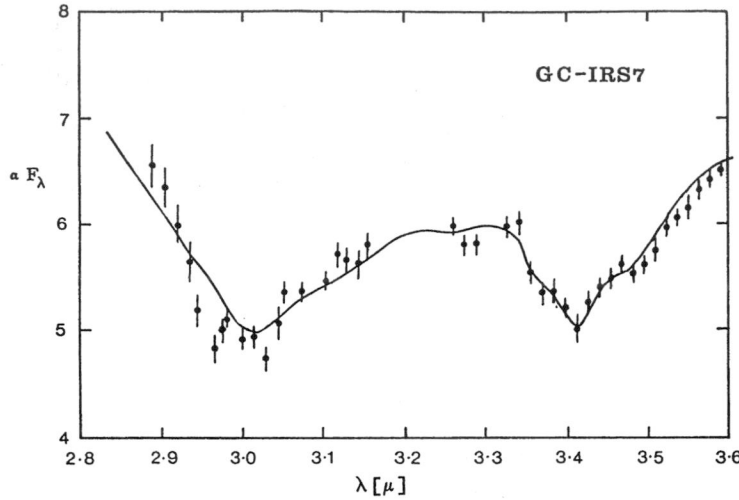

Fig. 2.2. Spectrum of an infrared source at the galactic centre (points) compared with the predictions of a bacterial (desiccated *E.coli*) model (curve).

so that whatever we say in our opposition will eventually come out on top in the argument. Experience shows that when there are no good observations in favour of what seems absurd this easily adopted policy is usually fairly safe. But in the face of good observations and in the face of many of them, as in Figures 2.1 and 2.2, it is a highly questionable strategy.

So just how good are the correspondences of Figures 2.1 and 2.2? By the early 1980's, when we attempted to answer this question, we had two decades of experience behind us in evaluating such correspondences. Expressed quite simply, we had never seen anything nearly so good. Yet even so was it all good enough to sustain a belief in such an apparently outlandish idea?

We recognised that to a person who had not followed the problem over the years, the absorption and scattering characteristics of the interstellar grains in these two figures might have seemed inadequate support for such a far-reaching hypothesis. And doubtless this was the way it appeared to many. But to us who had been involved over almost two decades it seemed otherwise, and we think it fair to add that time has supported our point of view here. Nobody among the critics of the 1980's has managed to find an alternative theory of the absorption characteristics of the grains to equal the success of the bacterial hypothesis.

There was a broad-ranging argument in this direction already available in the mid-1980's. The observations of Figure 2.2 show that the bulk of the grains must be organic. On this there can be no disagreement. Moreover the isotropy of Figure 2.1 shows that the grains must be substantially the same in one direction from the Earth as in another. Much the simplest way to produce a vast quantity (10^{40} grams) of small organic particles everywhere of the sizes of bacteria is from a bacterial template. It is not from any known form of abiotic process.

The immense power of bacterial replication is worth careful note at this point. Given appropriate conditions for replication, a typical doubling time for bacteria would be two to three hours. Continuing to supply nutrients, a single initial bacterium would generate some 2^{40} offspring in 4 days, yielding a culture with the size of a cube of sugar. Continuing for a further 4 days and the culture, now containing 2^{80} bacteria, would have the size of a village pond. Another 4 days and the resulting 2^{120} bacteria would have the scale of the Pacific Ocean. Yet another 4 days and the 2^{160} bacteria would be comparable in mass to a molecular cloud like the Orion Nebula shown in Plate 8. And 4 days more still for a total time since the beginning of 20 days, and the bacterial mass would be that of a million galaxies. No abiotic process remotely matches this replication power of a biological template. Once the immense quantity of organic material in the interstellar material had been appreciated, a biological origin for it becomes a necessary conclusion, in our opinion.

Now comes the question: Where did the interstellar particles come from? How did they get where we now observe them to be? This question leads us to another important step along the path, to the comets.

2.4 COMETS

An individual comet is a rather insubstantial object. But our solar system possesses so many of them, perhaps a few hundred billion of them, that in total mass they equal one of the outer planets, say a mass like Uranus or Neptune, about 10^{29} grams. If all the dwarf stars in our galaxy are similarly endowed with comets then the total mass of all the comets in our galaxy, with its 10^{11} dwarf stars, is 10^{11} $\times 10^{29} = 10^{40}$ grams, which is just the amount of all the interstellar particles. Check number one.

How would microorganisms inside comets get out of them? We know as a matter of fact that comets do eject particles, typically at a rate of a million or more tons a day. This was what Comet Halley was observed to do on March 30-31, 1986. And Comet Halley went on doing just that, expelling particles in great bursts, as long as it remained within observational range for the next ten years. The particles so ejected in March 1986 were well placed to be observed in some detail. No direct tests for a biological connection had been planned, however, but infrared observations showed them to be much like the interstellar particles responsible for Figure 2.2. It is clear that they were organic in composition, as have been the particles expelled from a series of new comets that have visited the inner solar system since 1986.

The radiation pressure from sunlight drives small particles expelled from comets rapidly outwards in the solar system, ultimately out into interstellar space. This is exactly what is happening when we observe the tails of comets. The tails consist of small particles, expelled rapidly outwards by sunlight. So this is how the particles responsible for Figures 2.1 and 2.2 get out into interstellar space. They come from a great multiplicity of events like that which occurred to Comet Halley on 30-31 March 1986, like all the cometary explosions that supply the materials of the comas[1] and tails of comets. Check number two.

Comets are believed to have formed already in the early stages of the condensation of the sun, an event which probably occurred within a gas cloud like that shown in Plate 9. Here we reach a delicate point in our argument. We require some small fraction of microorganisms present in the solar parent cloud to have retained viability, or to be capable of being reactivated after being incorporated inside newly condensed comets of the early solar system.

---

[1]The nebulous envelopes around the nuclei of comets.

The fraction could be exceedingly small. For one percent of the mass of the initial comet cloud being made up of interstellar dust the total number of 'graveyard bacteria' in a single comet would be some 10^{28}. A viable fraction as small as one part in 100,000,000,000,000,000 would still yield a hundred billion bacteria for each comet to start life with! Once replication starts inside a newly-formed comet, previous losses become irrelevant, because of the enormous capacity of even a single viable cell to increase its number.

But replication requires heating to some modest temperature, so as to produce liquid water. The cloud in which the solar system formed would be expected to have harboured one or more massive stars of the types producing the unstable isotope ^{26}Al with a half-life of 7.5×10^5 years. This is long enough for the solar nebula to form and condense before all the ^{26}Al is effectively gone. As ^{26}Al is an appreciable contributor to the stable isotope of magnesium, ^{26}Mg, which was a common element in the solar nebula, the original amount of ^{26}Al must then have been substantial. Compared to ^{27}Al, the main isotope of aluminium, ^{26}Al would be expected to comprise a mass fraction of about 10^{-4} downwards according to the time which elapsed since its stellar synthesis. This would give an energy release of 6×10^{10}ergs downwards per gram of cometary material. (The decay energy of ^{26}Al is about 6×10^{16}erg/gram.) But because only some 10^{10} ergs per gram would be needed to heat and melt the water content of a comet, it is clear that the energy supply from ^{26}Al is likely to be sufficient to do exactly that.

The likelihood therefore is that the early comets were liquid. This sets an interesting problem of elementary physics. After the exhaustion of the ^{26}Al in the comets, how would the large cooling sphere of water freeze? Not as a solid sphere of ice, for the reason that water has the remarkable property of expanding as it freezes. But as a highly fragmented object, a conglomerate of icebergs separated from each other by a hugely complicated system of cracks and crevices that fill eventually with substances of lower freezing points than water — the more volatile components of the comet. It is this complicated structure, coming increasingly apart under the influence of solar heating, that constitutes the observed activity of a comet, activity that is impossible for the dirty snowball paradigm, at any rate in anything like its original form.

Thus the comets are the likely places of replication of micro-organisms – or more accurately one of the important places of replication, leading to the general picture shown in Figure 2.3.

1.5. PANSPERMIA IN REAL TIME

The Earth is embedded in a halo of cometary material, derived in part from comets that visit the inner regions of the solar system only at intervals of thousands of years, and in part from short-period comets with orbits shown in Figure 2.4. It is estimated that upwards of a thousand tons of this halo material enters the Earth's atmosphere every year, sufficient when in particles of bacterial masses to provide for 1,000,000,000,000,000,000,000 (a thousand billion billion) particles. Or many more still if they are of the sizes of viruses. In this vast number one can wonder if some might retain viability. If so, how would they be expected to react with the Earth's present-day biosphere?

On reflection we felt that the best chance of being able to observe such an effect would be where the interaction happens to be of a pathogenic nature. This is the small fraction of cases where an incoming organism succeeds in establishing itself by replication, with some terrestrial plant or animal acting as a host. An explosive replication in the form of an epidemic attack of some new disease would give the most favourable possibility of observing the entry of microorganisms from the cometary halo into the Earth's biosphere.

There is a great deal of evidence to support the view that such attacks actually happen, sometimes in violent and spectacular forms. When this evidence is presented to orthodox biologists it tends to receive short shrift. But in our experience the evidence tends to be treated more seriously by medical doctors. Indeed from the beginning of our investigations in panspermia, our views have been consistently regarded with interest by the medical profession, particularly by those who are concerned with problems at a clinical level.

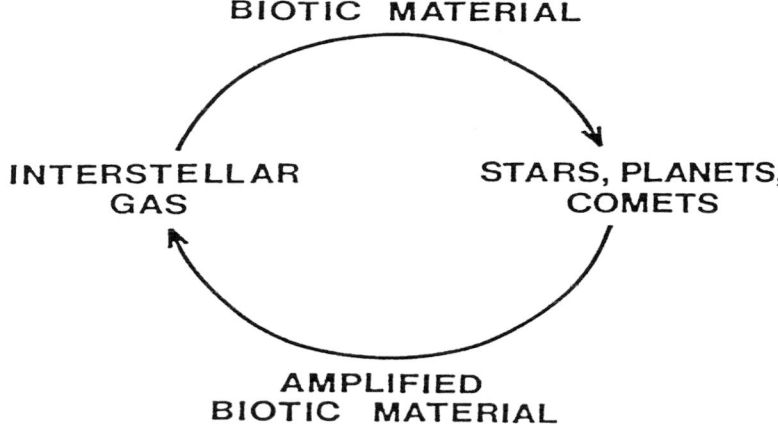

Fig. 2.3. Cosmic amplification cycle.

At this point it is worth looking at disease from the point of view of the survival of a pathogen. Inability to attack a suitable host obviously threatens extinction of the pathogen. But just as certainly, too much ability to attack a host threatens extinction, inevitably following the extinction of the host itself. Long-term survival of any pathogen would demand a perpetual walking of a tight-rope, not attacking too weakly and not attacking too strongly. Regarded in this way, it seems surprising that so many pathogens appear to have walked the tight-rope successfully, not just in our own day but throughout history. It is our opinion that in fact they do not and that no pathogen survives very long unless it is subject to repeated renewals from outside.

The lower the population density of the host, the more unlikely that a pathogen can maintain itself in the usually supposed way, by spreading from victim to victim. This is particularly necessary for rapidly acting pathogens, as many viral diseases are rapidly acting – measles for example. So how did the measles virus and other similarly violent pathogens maintain themselves among the relatively small human populations in prehistoric times? *Usually offered answer:* they must have existed then through attacking some other host, some host more common than humans.

In a few cases, doctors in ancient Greece gave remarkably good descriptions of the diseases of their day, the common cold for

Fig. 2.4. Projections of the orbits of known short period comets onto the ecliptic plane.

example. Why then is it that in most cases their descriptions are unrecognisable in terms of modern diseases? *Usually offered answer:* Ancient doctors were mostly inaccurate in their observations but occasionally, by luck, they got it right.

The first descriptions of modern diseases tend to be found from about the end of the 16th century onwards, for example the first description of influenza. Why were there no good descriptions before that? *Usually offered answer:* Before that people were unobservant.

Forest clearances in the Amazon valley and in other neighbouring valleys of South America have revealed hitherto unknown tribes, typically with total populations of about 500 people. One such tribe was found to contain a surviving victim of poliomyelitis. Where had the polio virus come from in such a very small group? *Answer:* Somebody with the disease brought it in from outside. *Question:* Sufferers from polio find mobility difficult. Even highly athletic people were stricken severely by it. So how did a sufferer manage to get into difficult terrain, requiring long and arduous journeys? *Answer:* They managed it somehow.

During the Black Death, the causative agent managed to get into extremely remote villages, where it often wiped out a considerable fraction of the population. These cases have been explained in the same way as the polio in the South American forests – as being caused through the arrival of a sufferer from outside. So why was it that outsiders did not manage similarly to import the disease into the cities of Milan, Liège and Nuremberg, which archivists tell us remained largely free of it. *Answer:* The archivists got it wrong.

Grey seal in Lake Baikal in Central Siberia were found a few years ago to be dying in considerable numbers from a respiratory virus. Some six months later seals in the North Sea were found to be dying of the same virus. How was this possible by victim to victim transference? *Answer:* A seal must somehow have managed to swim from the one to the other. *Objection:* Lake Baikal has no outlets to the sea. *Answer:* Then perhaps the virus was carried by a dog.

No reports have ever been made that at times of serious epidemics country districts are less afflicted than town districts. This was true long before country dwellers took to working in towns. How in a theory of victim to victim transmission was it possible for regions of low population density to be as badly affected as city regions of high density? *Answer:* City dwellers on contracting the disease must have rushed out into the countryside, the way polio sufferers in South America made their way into the distant forests.

The year 1918 is infamous in medical history for a devastating world-wide pandemic of influenza, which is said to have killed more people than all the battlefields of the First World War. In writing the history of the 1918 pandemic (*New England Journal of Medicine*, May 1976) L. Weinstein pointed out that there were several distinct waves of influenza in 1918, the first not having very serious effects. It was the second wave, caused presumably by a modified form of the influenza virus, that did the big damage. Of this second wave Weinstein wrote:

> 'The lethal second wave . . . involved almost the entire world over a very short time. . . . It was detected in Boston and Bombay on the same day . . . but took days to weeks before it reached New York City . . .'

How in 1918, when travel from Boston to Bombay took of the order of four weeks by boat, did a virus manage by person-to-person transmission to cross the thousands of miles between these two cities in a day? *Usually offered answer:* Weinstein got it wrong.

After investigating the outbreak of influenza in the island of Sardinia in 1948, F. Magrassi wrote:

> 'We were able to verify the appearance of influenza in shepherds, who were living for a long time alone, in open country, far from any inhabited centre; this occurred contemporaneously with the appearance of influenza in the nearest inhabited centres . . .'

How is this to be explained according to person-to-person transmission? *Usually offered answer:* The influenza virus was carried, not by people but by birds, and the birds shed their droppings on both shepherds and on the nearest inhabited centres. *Question:* Being hit by flying bird muck is not a particularly probable event. *Answer:* It is, if there are enough birds. *Question:* Did birds carry the virus from Boston to Bombay in 1918, all in a day? *Answer:* No, as already said, Weinstein got it wrong.

In the Spring of 1978 we collected information about the epidemic outbreak of influenza which occurred in schools during the winter 1977/78. The headmistress of a particular girl's school wrote to us with apologies that she could not send detailed numbers of cases. She said she had almost the whole school, with some 250 girls, all hit within a few hours of each other. How according to person-to-person transmission with an incubation period of two to three days was this essentially universal attack supposed to have happened? *Answer:* In that particular case the flock of birds was enormously large.

So it went on over more than a decade, with one disproof of the old ideas after another being brushed aside by orthodox opinion, and with answers that in many cases bordered on the ridiculous. Yet in every case there was an easy explanation in terms of a pathogen (virus, or bacterium or at least an activating component) falling to the Earth's surface from outside.

There are many more examples that could be given and to which we will return in Chapter 6. It is tempting to bow out of the present chapter with a caustic comment but we shall not, offering an optimistic one instead. That within only a short while of the panspermia (BFS) or *cosmicrobia* hypothesis being generally accepted, the progress in biology will be immense, enough perhaps for biotechnology companies to begin to show profits.

3
THE UNIVERSE
AND LIFE

3.1. INTRODUCTION

The so-called Copernican principle stems from discoveries in the 15th and 16th centuries AD that proved decisively that the Sun rather than the Earth was at the centre of our solar system. A generalisation of this point of view might imply rejection of any concept that assigned a privileged status to our particular corner of the Universe. It has recently come to be recognised, however, that a strict adherence to the Copernican principle may not always be necessary, and indeed that such an adherence could sometimes lead to difficulties. For example, the emergence of multicellular life-forms leading eventually to the evolution of humans, followed an unlikely accident of some kind that occurred scarcely half a billion years ago. The fact of our existence therefore makes the past half billion years of geological history strikingly different from the rest of geological time. Although the circumstances leading to the so called 'Cambrian explosion' of multicellular life may have been rare or exceptional, we have to accept this asymmetry in the geological record as an empirical fact; and if it tells us something about the Universe at large that would be an added bonus. Considerations of this general kind have given rise to a new guiding principle in science that has come to be known as the Anthropic Principle. In its strongest form this principle asserts that the Universe is the way it is *because* we are here to observe it, and so it may even be construed that the Universe had a purpose that was anthropocentric.

In the present context we interpret the anthropic principle in its more acceptable weaker form, the so-called weak anthropic principle to imply that the universe must be *consistent* with the existence of life, and in particular with the existence of human life. The surprise is that so much can be deduced, some things predictively, from a seemingly obvious statement. It is, for example, sufficient to break the physicist's concept of 'a typical observer'. If we can only exist in some special place or over some restricted time interval then that is where we must be, even if the chance of a randomly chosen abstract 'observer' lying in the spacetime volume in question happens to be small. The weak anthropic principle serves to remove otherwise inexplicable cosmic coincidences by the circumstance of our own existence.

One of the present writers was involved in an early application of the weak anthropic principle. Out of the 90 stable elements, 23 or so are necessary for life. There are four with high abundances in the cosmos – hydrogen, carbon, nitrogen and oxygen, nine with intermediate abundances – sodium, magnesium (in chlorophyll), silicon (in diatoms), phosphorus, sulphur, chlorine, potassium, calcium and iron, five trace elements that are present in organisms generally – manganese, cobalt, copper, zinc and molybdenum, and five more are present also in trace quantities in particular organisms – boron, vanadium, chromium, gallium and tungsten. Of these 23 elements only hydrogen is thought to be primordial. Hence the other 22 have had to be produced in some way. The case of carbon, the element whose complex chemistry forms the basis of life, is particularly interesting in this respect. It was shown in 1952-53 that to understand how carbon and oxygen could be produced in approximately equal abundances, as they are in living systems, it was necessary for the nucleus of ^{12}C to possess an excited state close to 7.65 MeV above ground level. No such state was known at the time of this deduction but a state at almost exactly the predicted excitation was found shortly thereafter. So one could say this was an example of using the weak anthropic principle in order to deduce the way the world must be, although the concept of the anthropic principle had not been explicitly formulated at that time.

Besides the need for the 7.65 MeV state in ^{12}C, a positive requirement, there was also a negative requirement. The nucleus of ^{16}O has an excited state at 7.12 MeV above ground level, which is just a little less than the sum of the rest mass energies of ^{12}C and an alpha particle (^4He). If things had been the opposite way, with the excited state in ^{16}O a little above $^{12}C + \alpha$, there would again have been no carbon in the world, because the nuclear reaction $^{12}C + \alpha$

$\rightarrow^{16}O$ would have taken out the carbon as fast as it was produced, there would have been a resonance as one says for the conversion of carbon to oxygen. So the approximately equal balance of carbon and oxygen in living organisms depended on the nuclei of these elements being rather finely-tuned in two respects, one in ^{12}C the other in ^{16}O.

When one examines the details of the situation more closely than we can do here, it is hard to avoid asking a more searching question: Is the favourable fine-tuning, favourable to life, just a matter of chance? Or is the situation in these nuclei somehow connected with the existence of life? If this were the sole grounds for asking this rather fantastic question we might feel inclined to dismiss it, as nowadays we would dismiss the chance coincidence that the angular diameter of the Moon is almost exactly the same as that of the Sun. But when one looks at other circumstances affecting the existence of life, for example in the details of the chemistry of carbon, and how these details depend on the numerical value of the so-called fine-structure constant, the same question arises repeatedly. The physical properties of matter appear to be adjusted to permit the existence of life. This form of words suggests a teleological (purposive) connection, which being unpopular in science has been replaced by the concept of the strong anthropic principle, according to which our existence somehow forces the physical properties of matter to take a form consistent with our existence. If the 'somehow' here could be satisfactorily explained in scientific terms, all would be well. But otherwise many will object, seeing the strong anthropic principle only as a semantic substitute for teleology, which by common consent is disbarred from science, because history shows the admission of teleology leads to fragmentation and disagreement in the way we look at the world.

In our own view, the value of a concept to science depends either on the predictability criterion or on the concept serving to tie together in a demonstrable way facts which hitherto had seemed disjoint. We ourselves have not seen how the strong anthropic principle can be tested in either of these respects, whereas the weak anthropic principle is indeed open to test, not just in regard to the example mentioned above, but in a far-reaching way in the subject of cosmology, which we shall consider in the present chapter. The discussion will proceed in four stages:

(i) an attempt to define the nature of life;

(ii) a determination of what might be called the information content of life;

(iii) a matching of the information content of life to that which various cosmologies might be expected to provide, with the inference that if a particular form of cosmology cannot match the information content of life then it is not the correct cosmology;

(iv) within a permissible cosmology astrophysical conditions must be arranged in such a manner as to permit the origin and evolution of terrestrial life.

Our discussion of the nature of life is intended to cover only the biochemical hardware of life. The neurological systems of higher animals can be thought of in terms of a computer analogy, with both hardware and software components. To many it seems as if the software component may have an existence independent of the hardware. The software may be considered to manifest itself with the phenomenon of consciousness, which is generally accepted nowadays by physicists to have a critical role to play in the interpretation of quantum mechanics. This adds considerably to the case of those who think there may be more to the software than straight-forward evolution involving hardware alone, a case which also touches on the strong anthropic principle.

Fascinating as such speculations may be, they do not form the topic of this chapter. Here we are concerned solely with the hardware of life and with what its information content may imply for cosmology.

3.2. THE NATURE OF LIFE

The atoms present in living systems are no different from similar atoms in non-living material. An atom of carbon in our bodies has the same individual physical properties as a carbon atom in a flake of soot. Yet the cooperative properties possessed by the arrangements of atoms in living matter are astonishingly different from those in inanimate material. You could store equal quantities of carbon dioxide and free hydrogen in a bell jar in the laboratory for an eternity and that is the way they would stay. But introduce a special kind of bacterium into the bell-jar and the gases will go in short order into methane and water. The bacteria in question are of a special kind which in recent years have become known as archaebacteria. They form a special kingdom, apparently without microbiological connections to other bacteria, or to the larger so-called eukaryotic cells of which ordinary plants and animals are built.

Defining the nature of life is one of those questions which becomes harder and harder the more you look into it. Instinct tells us that a snail is radically different from a stone. But why is it different?

Let us start an attempt to answer this question by noticing that the issue of which assembly of molecules is most stable (the proportions of their constituent atoms being specified) depends on the temperature. At laboratory temperature the most stable form for a suitable mixture of hydrogen, oxygen and carbon is methane and water. But at the temperature of a wood fire the most stable form is carbon dioxide and hydrogen. Add to this fact that mixtures of atoms do not necessarily reach their most stable forms. At higher temperatures like those present in the log fire they usually do, but at laboratory temperatures they may not. Start from methane and water in the laboratory and heat the mixture. Given adequate time it will go to hydrogen and carbon dioxide. Now cool the mixture. It will not return to methane and water, no matter how slowly you cool it. Unless of course archaebacteria happen to be present.

The most stable forms for mixtures with atoms of hydrogen, carbon, nitrogen and oxygen, the commonest atoms in living material, behave in exceedingly complex ways at laboratory temperatures, or at lower temperatures. But the most stable forms are generally not attained by inanimate mixtures. They are attained, however, or nearly attained, when living organisms are present. It is this property of being able to reach, via controllable means, the stable forms of mixtures at temperatures characteristically found on the Earth (300 degrees Kelvin) that defines the nature of life.

The substances on which living systems operate in this way are usually derived from mixtures of compounds that came to be assembled at higher temperatures. It is a general property that as mixtures go to their most stable forms with decreasing temperature, energy is released, not absorbed. Thus the ability of life forms to reach equilibrium states with lowering temperature provides them with energy sources; and it is on such sources that life in its simplest forms depends.

Science can be divided into its many branches by the magnitude of energy transitions. The biggest energy steps are those found in particle physics, running to thousands of millions of electron volts (eV). Accumulating basic information about such steps is difficult and consequently expensive. Most of the basic data on which theories in particle physics are based could be written on three sheets of paper, data which has cost billions of dollars to obtain. In contrast, basic data at energy steps of a few eV, obtained in the 19th and early years of this century, cost sums measured only in thousands of dollars. This was the data of atomic physics that led in its highest theoretical form to the development of quantum mechanics. Because of its history, the habit of thought in physics is to relate subtlety directly to energy, the larger the energy step the greater the measure of subtlety. Biology challenges this point of view. Biology says, conversely, that the lower the energy step the greater the measure of subtlety. This inversion of attitude is perhaps one of the reasons that physics and biology have become so sharply separated in our educational system.

The chemical bonds between atoms that have to be changed in reactions at low temperatures in order to achieve the most stable states are pretty much the same as in atomic physics, energy steps of a few electron volts. But whereas state changes in atomic physics are achieved by radiation units, quanta, with energies that are the same as those of the changes in question, quanta of a few eV, in biology the state changes are achieved with quanta of much lower energy, typically of about $1/40$ of an electron volt. This is done by exceedingly subtle accumulations of energy, by pumping through sequences of metastable states. An analogy might be to surmount a high wall step-by-step up the many rungs of a ladder instead of jumping over it directly. Or one might think of charging an electric battery over a lengthy period and of then discharging it in short order.

In one important respect biology also takes advantage of quanta with energies of about three electron volts, in the process of photosynthesis whereby carbon dioxide and water are reduced to oxygen and sugars, a similar result to the operation of archaebacteria but very different in its detailed operation. The operation is by no means completed through the higher energy quanta. Their absorption serves as an energy source which gives rise to a train of reactions of the more usual lower-energy type.

The substances that control the small energy steps of biology are proteins. Proteins consist characteristically of linked chains of amino acids of which 20 different kinds dominate the situation in biology. Only an exceedingly small fraction of the possible chains of amino acids are biologically relevant – just how small will form the main topic of the next section. Also characteristically, a biologically important protein (enzyme) will have a number of amino acids in its chain ranging from about 100 on the low side to about 1000 on the high side, with 300 as a fair average.

Although it is useful for diagrammatic purposes to think of a protein as a linear chain, enzymes actually take up amazingly complicated shapes in three dimensions, especially when suspended in water. The primary structure is a helix. Water is repellant to a fraction of the 20 amino acids and these, wherever they are in the chain, form a central region in the presence of water, so as to become shielded from the water by the others. This leads to a hugely complicated shape which is then given stability by chemical linkages, as for instance between the sulphur atoms that are present in just one of the 20 amino acids, the amino acid methionine. Such linkages are like the spars used in buildings to give strength to a human-made structure. Notice that although these so-called disulphide bonds occur between amino acids that are adjacent in space, such neighbours are not usually neighbours in the original chain. They have been brought together by the manner in which the original

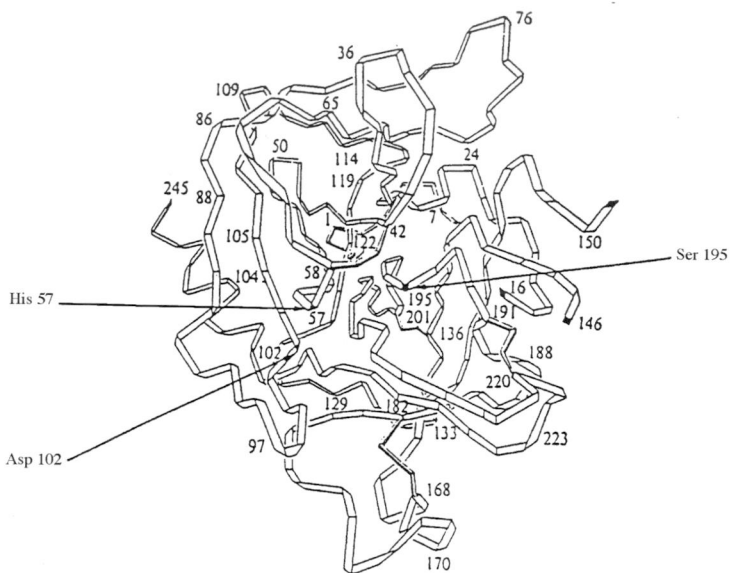

Fig. 3.1 Structure of the enzyme α-chemotrypsin. Three critical amino acids are marked with arrows. (Adapted from P.D. Boyer (ed.) Enzymes, 3rd ed., Vol. 3 (Academic Press, New York, 1971)).

chain has been folded by its water-repellant members. The extreme com-plexity of the situation is illustrated by a particular example in Figure 3.1. Determining a structure like Figure 3.1 is a difficult job for the experimentalist. So not unnaturally the experimentalist chooses the relatively simplest cases to study. Thus Figure 3.1 is towards the simpler end of the class of enzymes.

Enzymes do not have simple surfaces. On the outside they are irregular with one specially important cavity, the so-called active site. The shape of this cavity is crucial to making chemical reactions 'go' that would not 'go' under inanimate conditions, like those re-actions which promote the conversion of carbon dioxide and hydro-gen to methane and water in the case of the archaebacteria. What happens for a particular reaction is that the chemicals involved fit with startling precision into the cavity of the relevant enzyme, not just as pieces of a jig-saw fit, but in a specially reactive orientation with respect to each other. Moreover, the chemicals are jostled so as to promote the reaction by the amino acids with which they are in contact, the amino acids forming the active site. The jostling is not random. It is organised in the sense of the ladder-over-a-wall

Plate 10: The deepest ever view of the Universe obtained with the Hubble Space Telescope, showing a myriad galaxies, stretching over billions of light years. *(Courtesy, R. Williams, NASA)*.

Plate 11: Dust clouds near the Galactic centre towards the constellation of Sagittarius. (Courtesy Anglo–Australian Telescope Board).

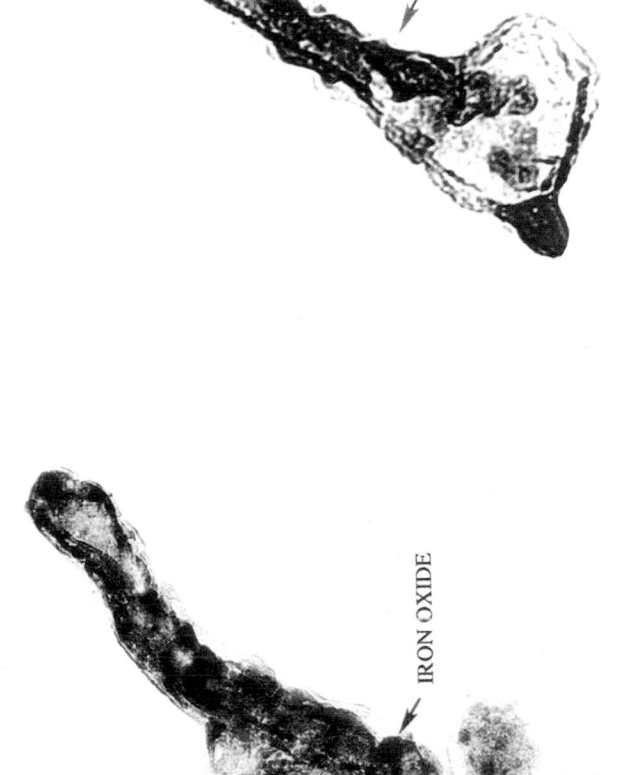

COMETARY PARTICLE

TERRESTRIAL BACTERIAL FOSSIL

IRON OXIDE

IRON OXIDE

0.1 μm

0.1 μm

Plate 12: A particle of complex organic composition collected in the Earth's upper atmosphere (J.P. Bradley *et al*) compared with a fossilised iron oxidising bacterium in the Earth's Precambrian sediments (H.D. Pflug). Note the similarity of structure and the iron oxide domains in the two cases.

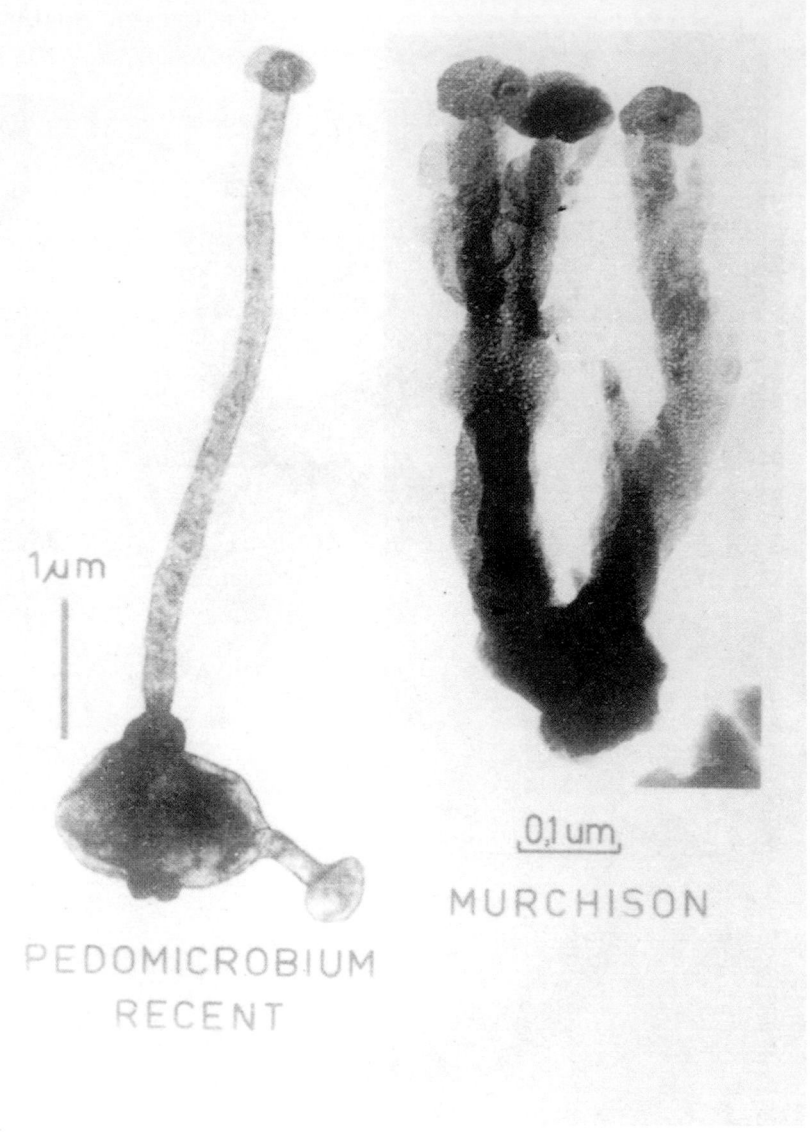

1μm

0,1 um

MURCHISON

PEDOMICROBIUM
RECENT

Plate 13: A structure resembling an iron oxidising microorganism in the Murchison meteorite compared with a similar present-day organism. *(Courtesy H.D. Pflug).*

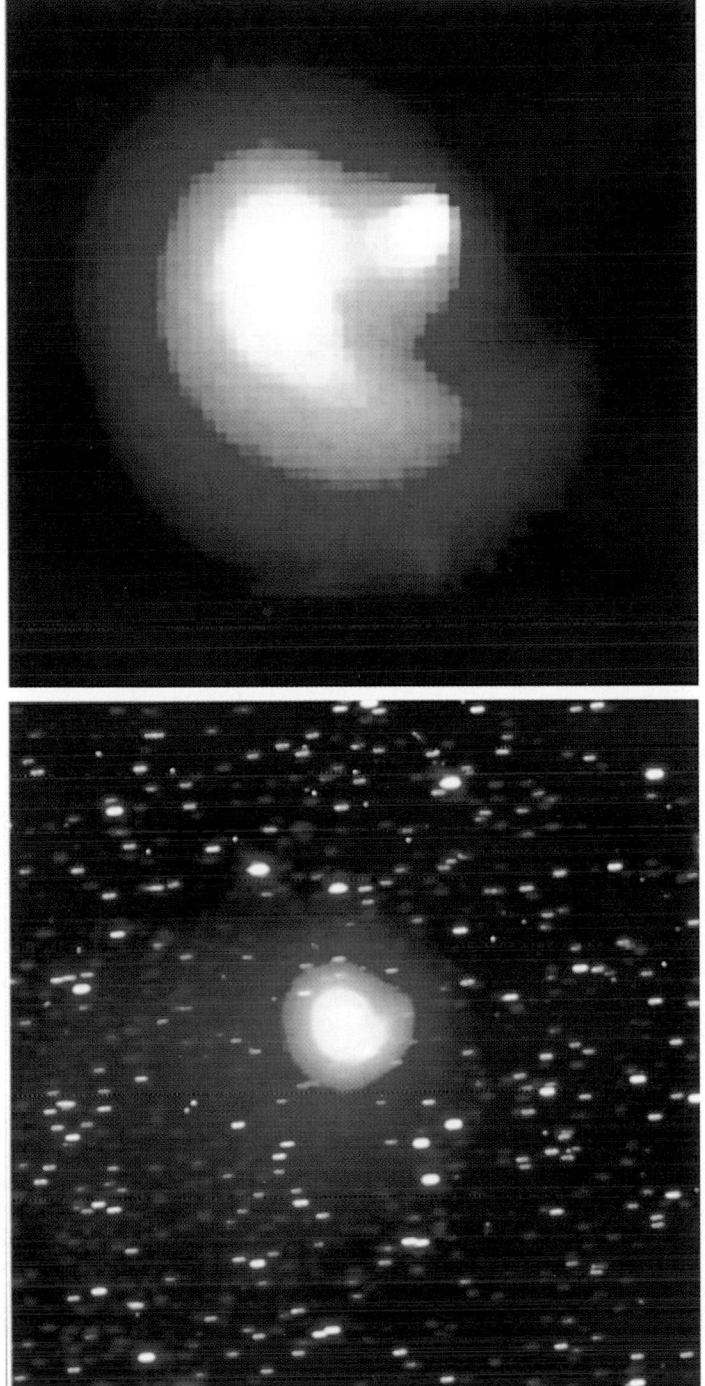

Plate 14: An Hubble Space Telescope image of Comet Hale-Bopp taken when the comet was in the cold depths of space, beyond the orbit of Jupiter. The close-up shown on the right-hand frame reveals a bright blob of reflecting dust. Cometary activity at this distance possibly indicates the effect of micro-organisms replicating beneath a hard-frozen icy crust. (*Courtesy H. Weaver and P. Feldman, NASA*).

Plate 15: Photograph of Comet Hale-Bopp taken on 13 March 1997 by A. Dimai, D. Ghirado and R. Volcan. The comet reached perihelion on 1 April 1997. *(Courtesy Sky and Telescope)*.

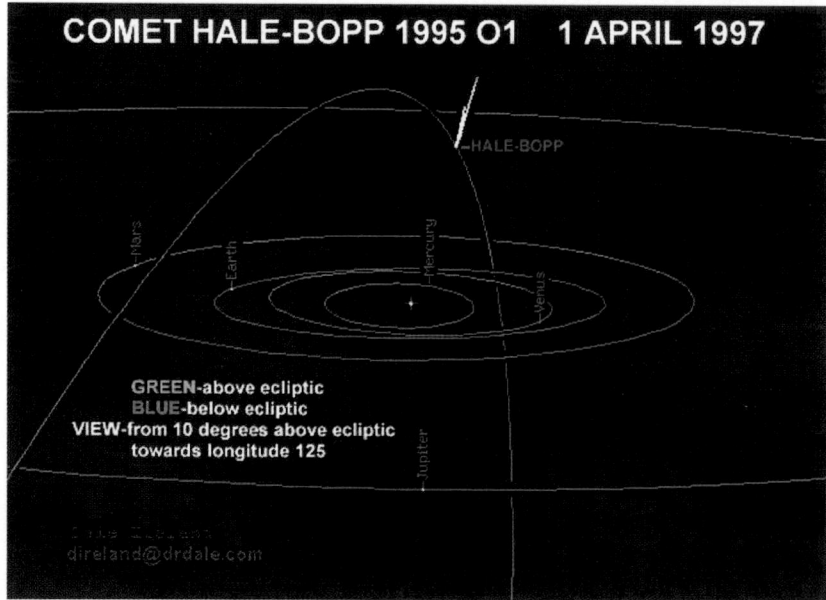

Plate 16: A view of the orbit of Comet Hale-Bopp in relation to the orbits of planets. The cometary orbit lies in a plane perpendicular to the plane of planetary orbits. *(Courtesy Dale Ireland)*.

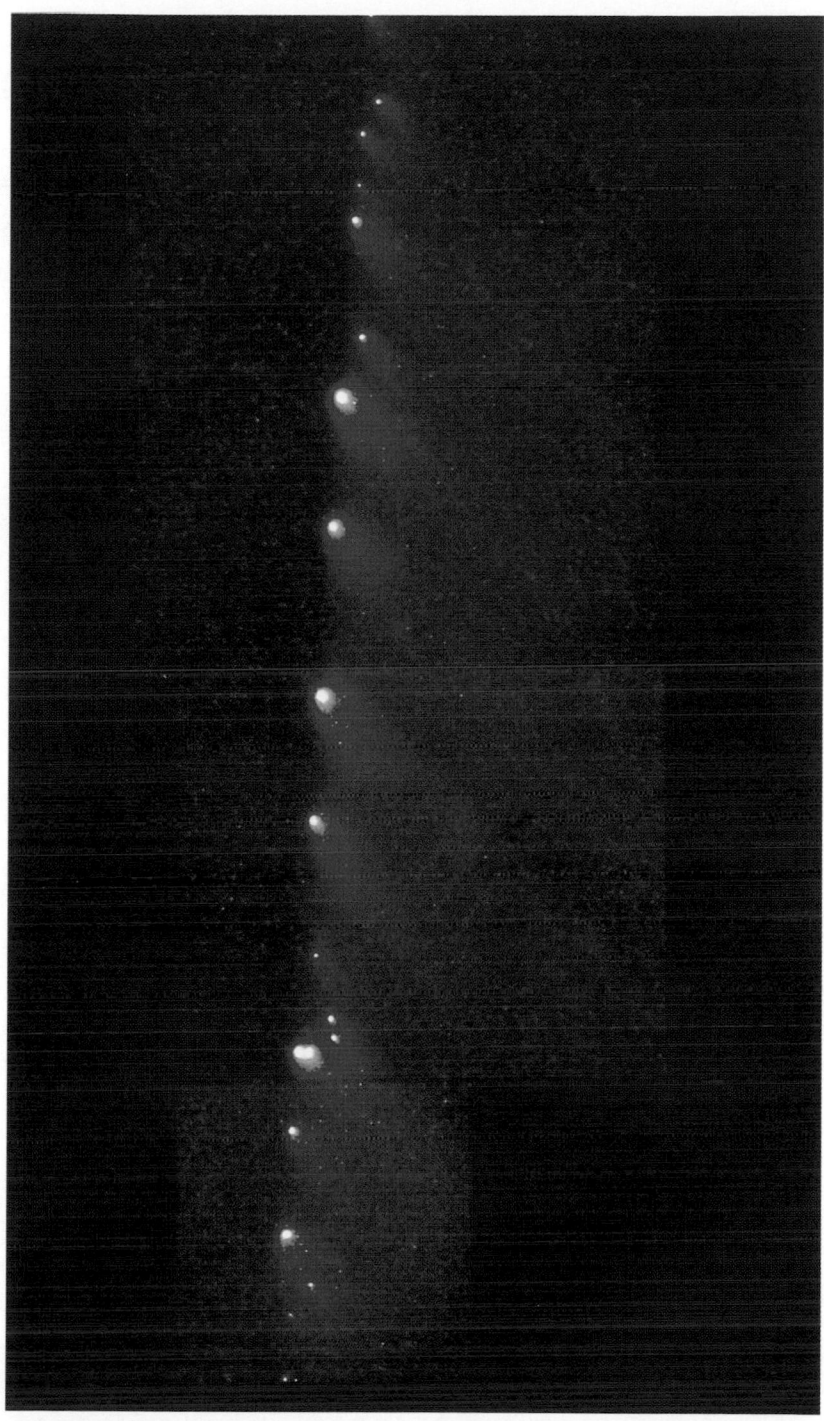

Plate 17: The fragments of the Comet Shoemaker-Levy that plunged into Jupiter in July 1994. *(Courtesy NASA)*.

Plate 18: The scene of devastation that followed the 1908 impact of a comet/asteroid fragment in Tunguska, Siberia. This photograph was taken in 1927.

analogy. When the reaction is completed, with the reacting chemicals having changed their shapes, they no longer fit the enzyme cavity as before. Consequently they break away from the cavity, freeing it to promote the same reaction yet again. And again and again in the manner of a catalyst. A catalyst is defined in chemistry as a substance which promotes a chemical reaction without itself being changed. Enzymes are catalysts analogous to human-made catalysts, but they are millions of times more effective.

Since an enzyme depends for its operation on a hugely precise matching of its shape to that of chemicals in a particular reaction, and since chemicals in different reactions have different shapes, enzymes are highly specific. Each promotes one particular reaction but not others, which is why living systems need many enzymes. The simplest living system needs 2000 or more enzymes, each matched to a particular reaction among the complex network required to sustain the system. In a complex system like ourselves, upwards of 100,000 highly specific amino acid chains are probably required to sustain the human network, although a precise count is hardly possible because of the network's immense complexity.

A living system has need of many copies of each of its enzymes. A literal accurate copying, amino acid-by-amino acid, of a structure like Figure 3.1 would be so difficult as to be hardly feasible. Just as we ourselves copy buildings from blueprints rather than by copying brick-by-brick or stone-by-stone, so copies in living systems are obtained from a blueprint. The blueprint is carried by four characteristic markers (nucleotides) read in blocks of three (codons) on the now-famous double-helix structure of DNA. The reading process is also vastly complicated. It is done mostly by the enzymes themselves. The first step is to construct an intermediate sequence of blueprints (the various forms of RNA). It is a case of the master blueprint of DNA producing through enzymic activity (not through its own activity – by itself DNA is very inactive) blueprint A, which then produces blueprint B, which produces blueprint C, ... until ultimately a considerably simplified and fragmented form is used to construct the enzyme in question. The raw materials for constructing the enzyme are separated amino acids which have to be linked together in the order prescribed by the eventual blue-

print. A similar logic is used in constructing a human-made building. The architect's drawings are more complicated than those
which are issued to individual workmen. But this type of human situation is simpler than the biological situation by a huge margin.

If one thinks there was a time before which life did not exist, a
conundrum arises in understanding its origin. Which came first, the
blueprint for an enzyme or the enzyme itself? If one says DNA
came first, the problem is that DNA is inactive. If one says the
enzymes came first, enzymes apparently cannot copy themselves.
The favoured answer among biologists is to say that an intermediate blueprint came first, a blueprint expressed by RNA not by
DNA. In recent years, RNA has been shown to possess a limited
degree of activity of its own, although whether the activity is sufficiently diverse as to be capable of maintaining a replicative system
still remains a question. The problem is one already hinted at above.
The bond strengths, whether in RNA or proteins, are in the region
of 4 eV, much too strong to be broken thermally. Thus a failure to
find a working system at the first joining of atoms stops there.
Without enzymes to break the bonds a second trial cannot be made,
except by flooding the material with so much energy that everything is smashed back into the constituent atoms. But such extreme
violence cannot lead anywhere, since floods of energy would also
destroy anything useful that might arise. There is but one way out
of this logical impasse, in our opinion, which is to make trials, not
repeatedly on a limited sample of material as in Darwin's 'warm
little pond', but to make just one trial on a breathtakingly large
number of samples. Just how large the number that would be
needed before anything interesting happened will be the topic of
the next section.

But this is not a situation with which we have much sympathy.
It is too remote from observation and experiment to be worth los-
ing sleep over. Indeed such experiments as have been done show
that, while it is not too hard to produce individual amino acids and
nucleotides from inorganic materials, no amount of human in-
genuity will persuade such products to arrange themselves in biologically interesting ways. Progress in this respect has been so
minuscule as to be essentially nil, which it would not have been if

matter had some hidden urge (as some mistakenly suppose) to arrange itself in ways suited to the origin of life. The evidence is that an origin, if such there ever was, turned on situations so unlikely that they cannot be rediscovered by chance in the laboratory.

Another reason for not worrying too much about the origin of life is that we have no knowledge or assurance that the problem is a real one. There may have been no origin, no time before which there was no life. Intuitively we may think there must have been, but if we do our instinctive supposition is cultural. It is not analogous to certain intuitive perceptions that lie at the base of mathematics, which everyone has regardless of culture. A Buddhist, for example, might think instinctively that life has always existed for an eternity in time.

Science does not make progress by searching out what appear to us subjectively to be the most important problems and by then hammering away in an attempt to solve them. Science makes progress by doing what happens to be accessible, by not wasting energy or resources on what is inaccessible at the moment. Accessible problems never depart very far from observation and experiment. There are many such issues with which we can be concerned without straying into vague speculation, issues of great interest with possibilities very different from orthodox positions, with lots of scope for the unusual. It is on such problems that we believe one should concentrate attention, as we shall attempt to do in the rest of this chapter.

3.3. THE INFORMATION CONTENT OF LIFE

With the invention of computers in the 1940's the idea of measuring the information content of a message was born, and a mathematical theory of how this might be done emerged to widespread applause from the scientific community. We never joined vigorously in the applause because the applicability of the mathematics seemed too restricted in its scope to be of much interest. What one would really like to be able to do would be to give a logical numerate meaning to the difference in the information content in the following two messages, supposed to reach the German Chancellory in Berlin on 1 June 1944:

> Message 1: *This morning the British Prime Minister, Winston Churchill, ate bacon and eggs for breakfast. Yesterday he smoked eleven cigars and sniffed brandy throughout the day. It is anticipated he will do the same on the 6th of the month.*

> Message 2: *Early on the 6th, the Allies will attempt to land very large forces on the Normandy beaches, from St. Germain in the west to Quistrehan in the east. There will be no landing in the Pas de Calais.*

The mathematical theory of information does not attempt to grapple with cases like these. Yet it is situations like these that are most important. Similar but still more awkward problems arise when the information content of life is at issue. Were a refined theory available for estimating the information content of DNA it would, in our opinion, be immediately apparent from its overwhelming content that life could never have arisen on a minuscule planet like on Earth. It would be seen that to match the information content of even the simplest cell nothing less than the resources of the entire Universe are needed. This is an opinion that can be backed up by making a shot at estimating the information content, noticing that if on reasonable grounds the answer turns out as vast beyond all precedent, it does not matter in its implications just how vast it really is, because one huge number would have the same implications as another. As a friend once put it:

> 'I wouldn't see much difference between inheriting £10 million and inheriting £1000 million. The effect on my life would be the same.'

For every enzyme needed to make a chemical reaction 'go' in the large complex of reactions that maintains a living cell, a number can be estimated in the following way. Take first the total number of proteins (and pseudo-proteins) that can be constructed by assembling at random the 20 biologically significant amino acids in chains of the same length as the enzyme in question, a length typically of some 300 amino acids. For such a length this number is unequivocal. It is about 10^{390}, i.e. 1 followed by 390 zeros. Next, divide this number by the number of possibilities in this set that serve to make the particular chemical reaction 'go' at an adequate speed to sustain the cell. Let us denote the average value of this latter number by the symbol f. Do this for every enzyme, 2000 in the case of a simple cell, 100,000 for a complex organism like ourselves. The result for the information content is then:

$(10^{390}/f)^{2000}$: simple cell

$(10^{390}/f)^{100,000}$: complex cell

The situation is still unequivocal. Scope for argument arises only when we come to estimate the likely average value of f. We saw in the previous section that an enzyme has to possess exceedingly specific properties in relation to the reaction which it catalyses. It has to curl up into a three dimensional structure with a surface cavity that provides a precise and special fit to the shape of the reacting chemicals. Moreover, the amino acids forming the cavity, the active site, have to be capable of jostling the reacting chemicals in a highly organised way. These properties depend crucially, not only on particular amino acids which form the active site, but on the positioning of the water-repellant amino acids which play a critical role in configuring the three dimensional structure. Another necessary property not mentioned in the preceding section is that an enzyme must be controllable. It must be capable of being switched on and switched off by chemical agents controlling the behaviour of a cell. Uncontrolled behaviour is what happens with cancers and this is to be avoided. Clearly all these drastic and precise requirements will not permit f to be unduly large, nothing like as large as the number 10^{390} appearing in the above formulae.

An extreme position would be to say that all these special requirements demand that the chain of amino acids be unique for each enzyme, demanding f = 1. This appears to be close to the truth in some cases. The protein histone-4 is found in both plants and animals and it has essentially the same amino acid structure in every organism. Little or no variants have been permitted throughout biological evolution. Human DNA has some thirty distinct genes

coding for histone-4. Variants are found among the thirty but they
are all of the kind that lead to the same chain of amino acids (same-
sense mutations). Other proteins are not as restrictive as histone-4,
however. But every enzyme that has been examined in detail has
been found to vary among plants and animals only to a moderate
degree. Summing up what has been found as fairly as we can, about
one third of the amino acids in a typical enzyme are obligate, which
is to say a particular amino acid must occupy each of about 100
positions in a chain of 300. The remaining 200 positions are by no
means free choices. Each of them can be occupied by three or four
among the bag of 20 amino acids, not by any member of the bag.
Arguing thus leads to $f = 4^{200} = 10^{120}$ (to sufficient accuracy) and
$10^{390}/f = 10^{270}$ giving the following for the information content:

$10^{540,000}$:   simple cell

$10^{27,000,000}$:   complex organism

These are not 'astronomical numbers', the description used popularly for large numbers. They are hugely greater than astronomical
numbers, the largest of which is obtained by dividing the distances
of the most remote galaxies, 10^{28} centimetres, by the scale of an
atomic nucleus. This yields the number 10^{40},

   10,000,000,000,000,000,000,000,000,000,000,000,000,000

when written out in full, certainly a big number, but nothing to
compare with the above numbers, which can be considered by
thinking how long one would need to write them out in full, and
how much paper would be used up in the process. Reckoning you
could write three zeros in every second, it would take only some 13
seconds to write out 10^{40}. But it would take nearly 7 months working 12 hours a day to write out $10^{27,000,000}$, and it would use up both
sides of some 10,000 sheets of paper.

   Evidently then, in measuring the information content of life,
we are dealing with *superastronomical* numbers on a grand scale.
Moreover, when one ponders over the unequivocal expression
$(10^{390}/f)^{100,000}$ it is clear that no reasonable choice for f can possibly
lead to anything other than a hugely superastronomical number.

Quibbling over the value of f will not lead to anything different. One superastronomical number is the same as any other in its significance, for it means that if we are to understand anything at all of the nature and origin of life we must search the universe for other superastronomical numbers. Only when we can match the superastronomical numbers from biology with a superastronomical number from cosmology can we expect to arrive at an insight into biology. Nothing could be more absurd than thinking that this can be done by contemplating events which have taken place only at the surface of the Earth. To imagine so is even less sensible than it was in days before Copernicus, when it was believed that the Earth was the centre of the Universe. The mode of thought is the same, but there is much less excuse for it today.

3.4. SUPERASTRONOMICAL NUMBERS FROM COSMOLOGY
How life could have evolved and spread throughout the Universe?

In this section we shall search for corresponding superastronomical numbers from cosmology. We begin by noting that with the exception of hydrogen all elements originate in stars, especially in supernovae. Thus stars provide the feedstock of life, just as they provide the inanimate materials of everyday life, the iron in the steel bodywork of a car for example. And we are all star dust!

The distribution of the elements is moderately uniform throughout our galaxy, and is believed to be much the same in most other galaxies. There is thus an approximately uniform distribution of the abundances of the elements throughout the universe. This cosmic distribution mirrors quite well the distribution of the life-forming elements, except that hydrogen is much more abundant cosmically than it is in living material. Carbon, nitrogen and oxygen are about ten times more abundant both cosmically and in life than the next group consisting of sodium, magnesium, silicon, phosphorus, sulphur, chlorine, potassium, calcium and iron, while the latter are about a thousand times more abundant than the trace elements. If one had to pick out an exception it would be phosphorus, which is some ten times more abundant in life than it is cosmically.

The complexity of the network of chemical reactions which define the nature of life depends crucially for its remarkable versatility on the properties of the carbon atom. Thus in estimating the quantities of potential life-forming material in various places within the universe, it is sufficient to specify the quantity of carbon, since the other elements follow along with the carbon in generally the required proportions. How these estimates go for a number of locales is shown in Table 3.1.

It is seen that superastronomical numbers appear in the second part of the table, but not in the first part. The meaning of the quantities in the second part is that if one starts with a chemical message (as for instance DNA is a chemical message) at a particular place at

Table 3.1

Place	Amount of carbonaceous material (grams)
Earth	10^{23}
Outer regions of Solar System (Uranus, Neptune, Comets)	10^{30}
Molecular cloud (e.g. Orion Nebula)	10^{35}
Interstellar material through our galaxy	10^{40}
All detectable galaxies	10^{50}

Limit for inter-related quantities of material in Big-Bang cosmology:

Time interval in Earth Ages (4.6×10^9) (years)	Mass of interrelated carbonaceous matter in Steady-State theories, (grams)
1	10^{50}
10	10^{59}
100	10^{140}
1000	10^{950}
1,000,000	$10^{900,000}$
100,000,000	$10^{90,000,000}$

a particular time, and if the message can be copied, then after the time intervals in the first column the message will have been spread by copying through the quantities of material in the second column. In the extreme case of the last line of the table, after a hundred million Earth-ages (4.6×10^{17} years) the message will be spread through $10^{90,000,000}$ grams of material, a number that is in a class which matches the biological superastronomical numbers of the preceding section. This suggests that life might be produced in a time interval of 10^{17} years provided the cosmology is steady-state.

Table 3.1 gives scope for a great deal of discussion. Here we shall simply indicate how the vast quantities of carbonaceous material in the second part of the table have been calculated. Biological cells typically have sizes of the order of one ten-thousandth of a centimetre, which happens to be just the size at which small particles are effectively repelled by the pressure of light, picking up speeds in the galaxy from starlight of several hundred kilometres per second. This is sufficient to spread a biological message everywhere through a galaxy in a time even less than a single Earth-age. It is indeed sufficient, just about, to spread the message from our galaxy to another, but only between neighbours. A still more powerful mode of spreading turns on the properties of iron as it is expelled from a supernova. When metallic vapours are cooled in the laboratory, condensation eventually occurs, not into more or less spherical globules, but into threads or 'whiskers'. Diameters of whiskers are typically about a millionth of a centimetre and lengths typically about a millimetre, giving the very large ratio of about 100,000 for the length to diameter. Such metallic particles are extremely strongly repelled by radiation in the far infrared region of the spectrum, and since molecular clouds in galaxies emit radiation strongly in the far infrared, whiskers can be repelled from galaxies into extragalactic space at speeds upwards of ten thousand kilometres per second, when distant galaxies can be reached from the galaxy of their origin in only a single Earth age. About a million galaxies can be reached in this way. Whiskers from millions of galaxies thus mix, producing a very uniform distribution for iron whiskers in extragalactic space, as for instance in the space between the galaxies in the deep HST field of Plate 10.

Of course iron carries no biological message in itself. But contiguous particles in a near vacuum have a marked tendency to stick together. A carbonaceous particle carrying a biological message could quite well stick to an iron whisker, hitch-hiking a lift across extragalactic space. One is reminded of the story of how the birds, after quarrelling as to who among them should be King, decided that it should be the one that in a trial was able to fly highest. Each kind fell back in the trial, leaving the eagle eventually to soar above the others. Yet even the eagle at last reached the height of exhaustion. When it did so, the wren, which had so far travelled unnoticed on the eagle's back, took off and with an effort attained a few feet more. So it came about that the wren became the King of birds.

In the present context it should be noted that any individual biological cell exposed to stellar ultraviolet and cosmic radiation over millions of years would suffer loss of viability. It is relevant, how-ever, that in all the laboratory studies available to date, it appears that for a whole bacterial culture that is subject to irradiation the loss of viability is never a full 100%. In the interstellar situation discussed above the minutest fraction of survival would suffice to transfer biological information derived in one location across astronomical or even cosmological distances.

The expansion of the universe does the rest of the spreading of the message. After reaching some million galaxies in the first Earth-age, the expansion approximately doubles the radius of the cosmological region containing the message for every succeeding Earth-age. After a million Earth ages (4.7×10^{15} years) the radius of the region therefore increases by $2^{1,000,000}$, and the amount of material in the region increases by $2^{3,000,000}$. In the latter connection it will be recalled that the essential difference between Big Bang cosmology and steady-state cosmology is that the universe does not empty as it expands in the Steady State case. It is this critical property of Steady State cosmology that leads to the vast quantities of material in the second part of Table 3.1, quantities that match biological requirements.

There is no reason why the standard qualitative picture in biology of the origin and evolution of life should not be given expression in this way. But it must be given expression in a cosmological setting, and the cosmology must be either steady-state or one of the more complex cosmological models that possess the basic steady-state property of self-replication. It cannot be a one-shot affair like the Big Bang. These are startling conclusions on which a great deal of evidence can be brought to bear. But for the present let us conclude by mentioning another way of arriving at significant superastronomical numbers. Start with a single living cell, say a bacterium. Let us now repeat an argument that we already discussed in Chapter 2. A typical doubling time by binary-fission for a bacterium supplied with appropriate nutrients would be two or three hours. Continuing to supply materials, the initial bacterium would generate some 2^{40} bacteria in 4 days, yielding a culture of the size of a cube of sugar. Continuing for a further 4 days and the culture, now containing 2^{80} bacteria, would have the size of a village pond. Another 4 days and the resulting 10^{120} bacteria would have the scale of the Pacific Ocean. Yet another 4 days and the 10^{160} bacteria would in mass be comparable to a molecular cloud like the Orion Nebula, and another 4 days, bringing the total time interval to only 20 days, and the mass of bacteria would be that of a million or more galaxies. In a year there would be some 2^{3650} bacteria and in a thousand years the total would be $2^{3,650,000}$ bacteria. Thus biology yields superastronomical numbers as well as depending on them.

Nutrients could not be continuously supplied it might be objected. Yet, cosmically speaking, the situation is nearer to a continuous supply than one might at first think. Formaldehyde (COH_2) is built as a weakly-bound molecule from the two commonest molecules in the universe, carbon monoxide (CO) and hydrogen (H_2). Although formaldehyde is not itself a substance of surpassing interest, take five or six formaldehyde molecules, swop atoms a little from one to another and join them appropriately, and you have all the sugars, the driving foodstuff of biology. Eliminate a water molecule between sugars and you have all the carbohydrates. Join sugars through nitrogen atoms and you have materials like the shells and claws of prawns and lobsters. A continuous supply is pretty well what one really does have in fact. It is rather here on the Earth where supply is limited, not in the Universe at large.

Let us now draw together what can be said from the above

considerations. The numbers in Table 3.1 show that once a replicative system emerges in the steady-state theory, a replicative system of any kind, it will spread throughout volumes of space and quantities of material that increase exponentially with time. Life need not have arisen all in one go. There could be a sequence of steps A,B,C, . . . with the evolving associations AB, ABC, ABCD, . . , one step being piled on another in an evolutionary process, again provided the reproducibility criterion is satisfied at all stages. Then the probability of life arising is of the order of the *sum* of the probabilities for each of A,B,C, . . . taken separately. For Big Bang cosmology, on the other hand, because of the limited time scale, which prevents spreading exponentially to superastronomical numbers, the prob-ability of life arising is of the order of the *product* for each of A,B,C... taken separately, a superastronomically smaller probability than when the sum rather than the product is taken.

3.5. DISTRIBUTION OF LIFE IN GALAXIES

We have already seen that the existence of terrestrial life requires a mechanism for the production of its constituent elements in the correct proportions. By 'correct' in this context is meant 'appropriate for the emergence of life'. This requirement in turn led to the prediction by one of us of an hitherto undiscovered excited state in the nucleus ^{12}C. The prediction was later verified by experiment, thus paving the way to a better understanding of the synthesis of metallic elements in stars. The probability arguments developed in earlier sections of this chapter relating to the problem of the initial origin of life will be shown now to imply further predictions from the weak anthropic principle.

The complex molecular components of terrestrial life must of necessity be distributed in the Universe in such a way as to be accessible for assembly here on the Earth. Astrophysical conditions in the early solar system would have to be so devised as to incorporate these components and to transfer them in a viable form onto the surface of the Earth. The molecular components must not only be of a complex organic character, they must also effectively contain the basic information of life to which we have already referred. The latter requirement, in our view, logically forces an identification with externally derived genetic structures as are present, for instance, in bacteria and viruses.

From such a standpoint we found ourselves able to predict the presence of complex organic polymers in interstellar space long before conclusive evidence for their existence became available. We predicted an infrared spectroscopic signature of cosmic microorganisms that was subsequently observed in both interstellar and cometary dust, confirming once again an expectation based on the weak anthropic principle. We shall describe these developments in the concluding part of this chapter.

The story begins in 1961 when the present authors set out to investigate the properties of interstellar dust grains in the galaxy. The millions of clouds of diffuse material that populate the Milky Way are filled with such particles, which show up as a cosmic fog, dense enough in many instances to blot out the light of distant stars. An example of such interstellar dust clouds is to be seen in the dark patches of Plate 11 which represent clouds near the centre of our galaxy. When we began our work on this subject astronomers, almost without exception, believed the dust to be composed mainly of volatile materials such as water-ice. On the basis of our own calculations we argued strongly against a water-ice composition and favoured instead a composition based on the element C, possibly occurring in the form of graphite flakes. As time passed and further astronomical evidence accumulated, it became clear that the situation was more complex than we had originally thought. By 1974 we found it necessary to include complex organic polymers in the cosmic dust composition, and from 1977 onwards we argued in favour of biopolymers similar to cellulose.

In our view a decisive verification of bacterial model came when the galactic centre infrared source GC-IRS7 was observed by D. T. Wickramasinghe and D. A. Allen (*Nature*, 294, 239, 1981) over the waveband 2.9-4µm with results as shown in Fig. 3.2. The curve in the upper panel is calculated for bacterial particles in the line of sight to this source, using the measured laboratory spectrum of desiccated *E. coli* obtained by S. Al-Mufti and shown in the lower panel. The agreement with the data points is clearly impressive. The bacterial grain model unquestionably leads to a close fit to the astronomical data. But could the logic be inverted to infer a bacterial model from the closeness of the fit? The issue inevitably hinges on cultural constraints as we have already mentioned in Chapter 2. The question arises as to whether the identification implied in Fig. 3.2 can be considered essentially unique. Could a 'simpler', culturally more acceptable non-biological system be invented to reproduce the same results? Although one has to admit that the shape of this absorption profile might possibly be reproduced by some synthetic distribution of suitably chosen C-H carrying structures, such a mixture has not yet been discovered, with the possible exception of organic residues extracted from irradiated ices to which we have referred in Chapter 2.

We consider a biological interpretation of these results to be overwhelmingly simple as compared with any *ad hoc* assignment of the same spectral features to an abiotically synthesised organic mixture. For abiotic solutions one has to ask the question as to how one could get the high degree of efficiency of organic production that is demanded by the astronomical observations. To get a large fraction of all the interstellar carbon tied up in the form of complex organics that resemble biomaterial, by means of non-biological processes, cannot be easy to say the least. It hardly needs to be stressed that the efficiency of biology in generating organic material cannot be matched. Well over 99.99% of all the organics found on the Earth results from the operation of biology! With the rapid march of modern technology, Man would have often wished he could make petroleum to fuel his cars and aeroplanes with a corresponding efficiency, but he has not succeeded as yet, and probably never will.

Another observation that over many years may have pointed in the direction of cosmic biology is the list of interstellar molecules as shown in Table 2.1 (p.28). There is now a vast array of carbon based organic molecules found in interstellar clouds which astronomers are valiantly trying to understand in terms of purely abiotic formation processes. This list is constantly being added to, the latest additions being the biological amino acid glycine, and vinegar (acetic acid). To explain the molecules of Table 2.1, as well as to understand

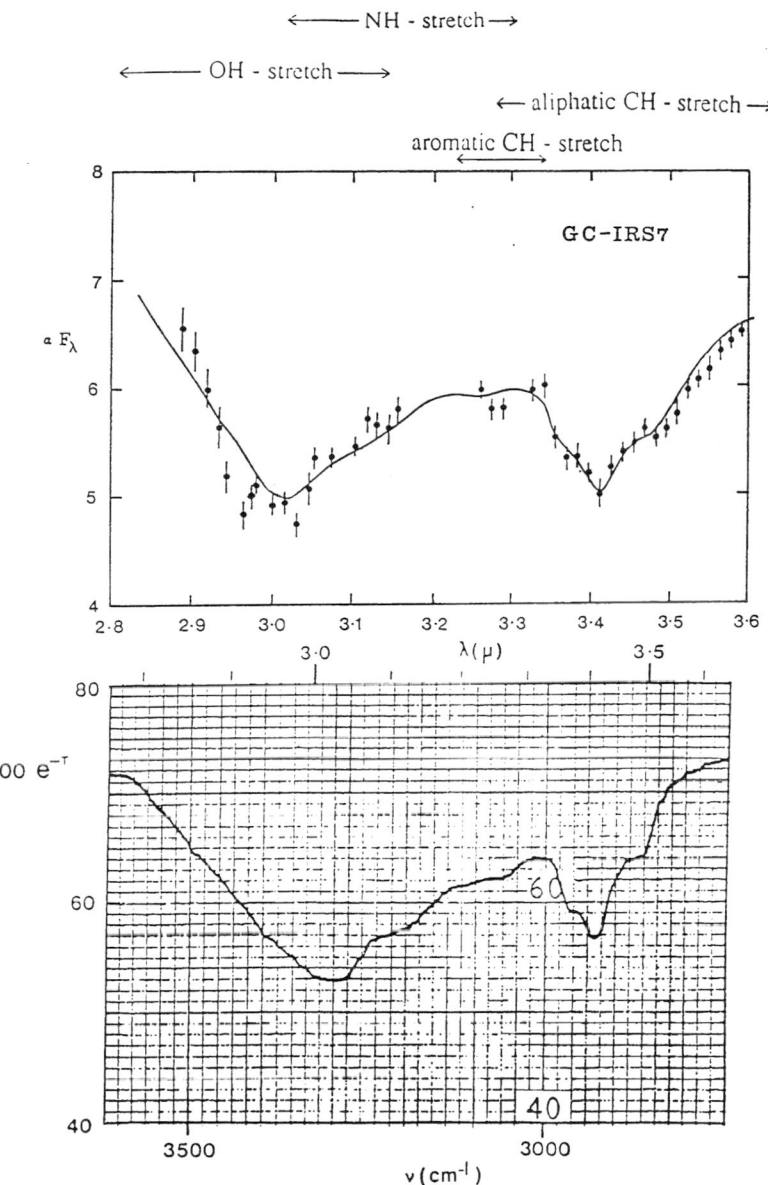

Fig. 3.2. *Upper panel:* Spectrum of an infrared source at the galactic centre (points) compared with the predictions of a bacterial model.
Lower panel: Calibrated transmittance curve of dessicated *E.coli* from work of S. Al-Mufti.

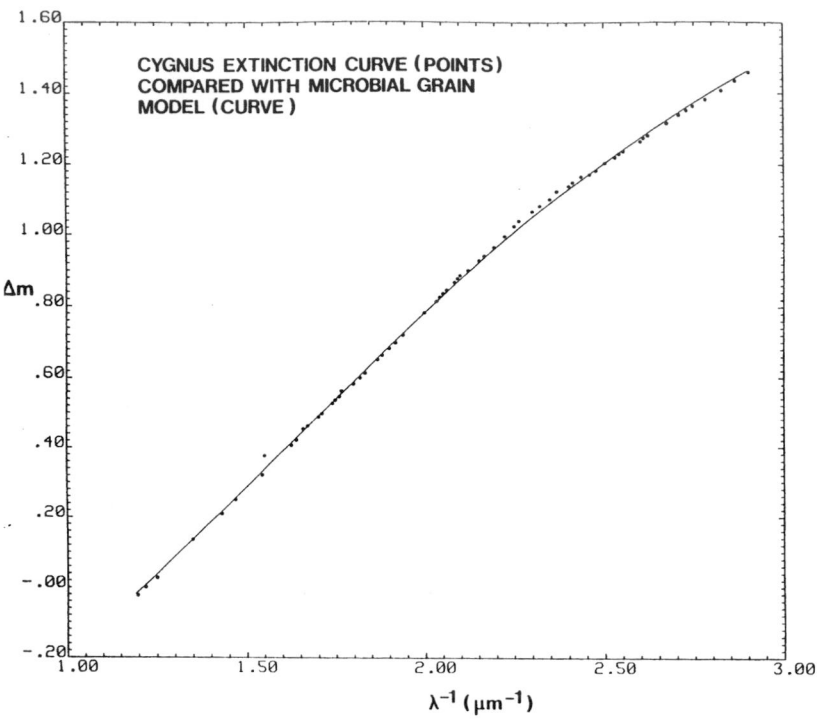

Fig. 3.3 Hollow bacteria and the visual extinction curve from the observations of K. Nandy.

the production of organic residues capable of explaining spectra such as is shown in Fig. 3.2, operationally at least, the simplest model is biology. A molecule related to the biochemical chlorophyll has for many years, been known to account for otherwise inexplicable absorption bands that appear in the spectra of stars – the so-called diffuse interstellar bands. Similar biochemicals have also been discussed in relation to an interstellar fluorescence phenomenon over the 6000-8000A waveband that has recently been discovered. Together with another group of interstellar molecules, the poly-aromatic hydrocarbons, that have been recognised to exist from observations of their infrared emissions at 3.28 micrometres as well as absorption at 2200A, the molecules in space may be seen to represent a graveyard of decaying interstellar bacteria – the by-products of the degradation of bacterial cells.

　　　There are other pointers to interstellar microbes as well. We know that starlight is dimmed as it traverses interstellar dust

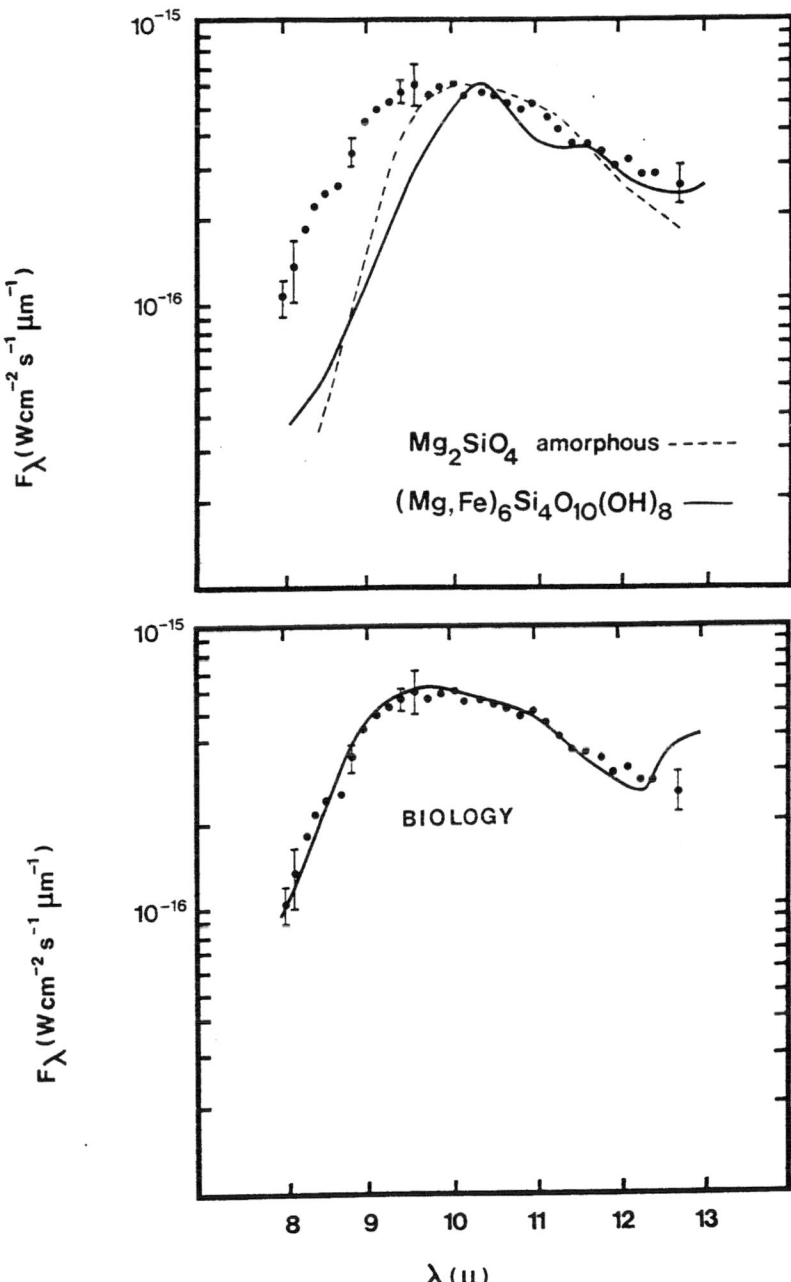

Fig. 3.4 Observations of heated dust in the Trapezium nebula (points), compared with emission curves of silicates and biological material (diatoms + *E.coli*).

clouds, and the extent of dimming varies with colour (or wavelength) in the manner shown by the points in Fig. 3.3. (See also Fig. 2.1 on p.31). The solid curve represents the calculated behaviour of a model involving the scattering of light by hollow bacterial particles. The agreement is seen to be good, pointing strongly to the correctness of the model. Amongst other evidence that pointed to a contribution from organic grains was the 8-14 micrometre spectrum of dust in the Trapezium nebula. Fig. 3.4 shows a comparison of the Trapezium nebula data with inorganic silicate models (upper panel) and with a combination of siliceous and organic matter as represented in a group of microorganisms known as the diatoms (lower panel). From Fig. 3.4 it is seen that silicate material on its own is ruled out, whereas a biogenic organic-siliceous mixture leads to an excellent fit.

The correspondences we have chosen for discussion in this section are only a selection from a larger number, and in our view they provide strong evidence that cosmic dust is largely microbial in character. On the basis of such comparisons we can estimate that throughout the Galaxy a total mass of 10^{33} tonnes of microorganisms appears to exist in a freeze-dried state at a temperature of about 10-25 degrees above absolute zero. The power of sequential doublings of bacterial cells that we discussed earlier would seem to be in evidence on a vast galactic scale. The cosmic amplification cycle of biology in this picture is summarised in Fig.3.5.

A rapid transformation of inorganic material into biosubstances would take place during the formation of new stars, planets and comets. Conditions in the outer regions of a nebula associated with a new-born star would allow liquid water and organic nutrients to be present within a multitude of comet-sized objects for millions of years. In the interiors of such objects cosmic biology can effectively explode. A fraction of the newly-formed biological particles are inevitably thrown out into space, the expulsion taking place by the action of radiation pressure from stars – starlight propelling the particles at high speed.

A newly-produced crop of bacterial grains is thus sent to join myriads of other cells which await them in a state of freeze-dried dormancy in the dark clouds of space. A new generation of stars and

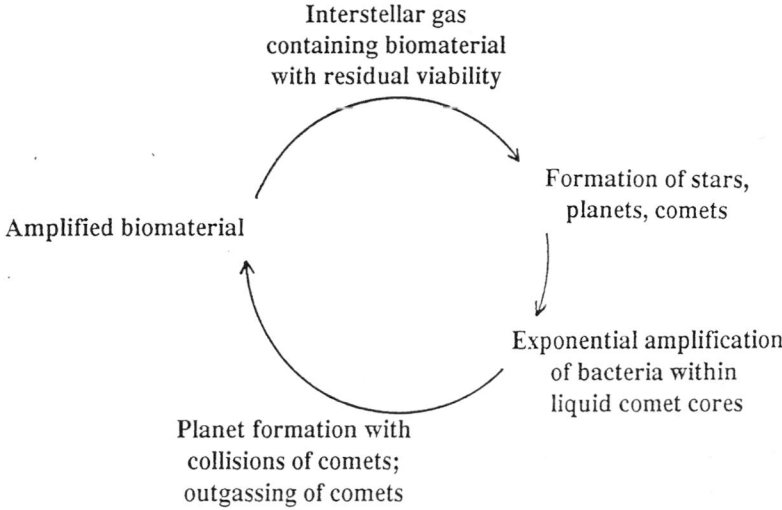

Fig. 3.5 Positive feedback loop for microbiology.

star clusters then spawn from these clouds and the whole cycle of cosmic biology continues with a positive feedback of biotic material into space. At the present time in our galaxy alone, approximately 10^{10} cycles in this loop have occurred, one for every sun-like star. The existence of dust is by no means confined to our own galaxy. Dust lanes are a common occurrence in external galaxies – in spiral as well as irregular galaxies – and there is no evidence to suggest either a composition or mode of origin any different to the dust in our own galaxy. On the contrary, wherever detailed spectroscopic evidence is available, close similarities to galactic dust show clearly, as for instance in the Magellanic clouds and the starburst galaxy M82.

Multiplication of life in comets

Our immediate biochemical antecedents are connected not with very distant galactic locations, however, but with situations that can be recognised as being relatively local. The nearest amplification sites of cosmic life, in our view, must be found within comets – objects which measure some tens of kilometres across and with whose debris the Earth frequently interacts.

At the birth of the solar system, from its parent molecular cloud, cometary bodies condensed at about the distances of the present planets Uranus and Neptune. At this time the interiors of comets were liquid due to the release of radioactive energy. We can argue that even the smallest population of cosmic bacteria present at the time of formation of this cometary cloud would have been vastly amplified within comets in their warm watery interiors on a very short timescale. When they subsequently cooled to become hard frozen, biology was trapped within them.

Comets become gradually stripped of volatile materials when they approach the inner regions of the solar system. Biological material lying within them may thus be progressively peeled away. Some of this particulate material rains down intact onto the surfaces of the inner planets. Some of it is splintered into atomic and molecular fragments by the action of solar radiation, the rest is expelled intact into interstellar space.

With the perihelion passage of Comet Halley in 1986 came yet another opportunity to test predictions of cosmicrobia (panspermia). We had now to show that the comet spewed out organic particles of a type that are indistinguisable from bacteria. It was found, both from space-probes and from ground-based studies, that predictions of this model were verified, as far as they could be. The general expectation of astronomers to find a dirty snowball comet, shedding particles of ice or mineral dust proved grossly wrong. The dust from comet Halley *was* found to be largely organic, and 90 percent of the comet's surface *was* covered in a dark inert crust, exactly as we ourselves had predicted months ahead of the encounter of the spacecraft *Giotto* with the comet. The infrared spectrum of Comet Halley showed again the signature of bacteria as displayed in Fig.3.6. Furthermore, *in situ* studies of comet dust by J. Kissel and F.R. Krueger (*Nature*, **326**, 760, 1987) using mass spectrometers aboard the *Vega* and *Giotto* spacecraft showed that the dust particles, as they crashed onto the detector and fractured, were made up of all the basic organic chemical groups that are usually associated with life. Although a possible connection with life was rejected by these authors on the grounds that a strong peak in the element phosphorus (P^+) was not found, this objection was later

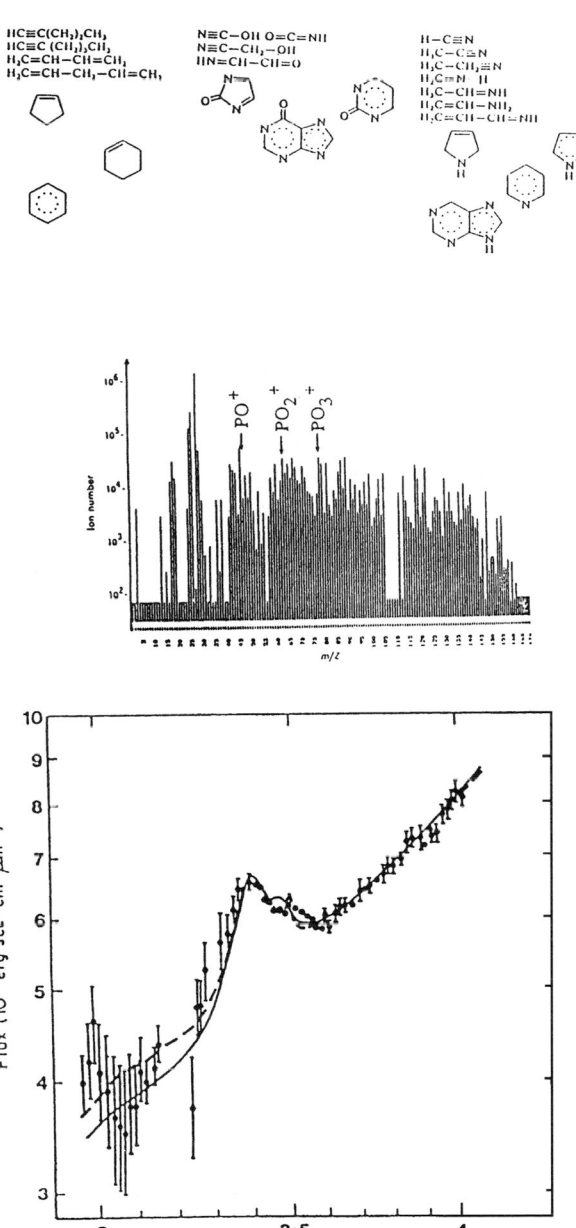

Fig. 3.6. *Top:* Distribution of mass in molecular fragments, leading to compositional groups as indicated *(adapted from J. Kissel and F.R. Krueger).*
Bottom: The infrared spectrum of dust from Comet Haley on March 30, 1986 obtained by D.T. Wickramsinghe and D.A. Allen (points), compared with predicted curves for bacterial model.

shown to be spurious. The break-up of phosphate groups, as occur in living matter, would be expected to yield not P^+ but the ions PO^+, PO_2^+, PO_3^+, for which strong peaks were indeed found (see N.C. Wickramasinghe, in *Infrared Astronomy* eds A. Mampaso et al, Cambridge University Press, 1993).

Normally in science, the fit of a prediction from a model to an observation like that displayed in Fig. 3.6 would be regarded as strong evidence in its favour. Yet the cultural bias against cosmicrobia (panspermia) was so intense in 1986 that astronomers were still making great efforts to search for alternative non-biological models, but without any real success.

What came as an even greater surprise in later explorations of Comet Halley was that a prodigious output of organic particles continued sporadically even when the comet had retreated beyond the orbit of Jupiter. A dust halo was observed when the comet was half way between the orbits of Saturn and Uranus. Millions of tonnes of organic dust continued to be spewed out from the nucleus of Halley, baffling the vast number of astronomers who still clung tenaciously to the idea of a lifeless dirty snowball comet. Biological activity building up sub-surface high pressure pockets that periodically explode would seem to provide a natural explanation of all these facts. A very similar pattern of behaviour has been repeated in recent years for other comets, notably in the case of the newly discovered comet Hale-Bopp, which had already developed an oblong coma of dust at the radius of Jupiter's orbit. These observations alone would be sufficient to dispose of the 'dirty snowball' model of comets once and for all, for snowballs, dirty or otherwise, do not explode in the cold depths of space.

Another property of comets, untypical of 'dirty snowflake' models, was observed in the Spring of 1996 for the case of Comet Hyakutake. A German-American team discovered that the coma of this comet, during the period March 26-28, 1996, emitted a hundred times more energy in soft x-rays than was expected. The x-ray images were obtained by the x-ray astronomy satellite ROSAT as the comet came into its field of view. This discovery has caused puzzlement among astronomers, but a simple explanation could be offered. The sun's corona is a strong emitter of x-rays and these x-rays could have been scattered into the field of view of ROSAT provided a sufficient quantity of very small carbonaceous particles were present in the comet's coma. We calculate that a few tenths of a megatonne of 30 Angstrom radius (virus-sized) carbonaceous particles are all that is needed to do the job. A total mass of dust of this order, large though it might sound, is no surprise because Comet Halley was found to be spewing out more than a megatonne in a single day in March 1986. The surprise, however, is that the new x-ray data seems to establish beyond doubt the existence of small carbonaceous, virus-like particles in a comet. This discovery could have far-reaching consequences in furthering our understanding of the nature of cometary dust. In particular it is fully consistent with the idea that viral particles might emanate from comets.[1]

At the present time the entire solar system is surrounded by a spherical halo of comets numbering some hundreds of billions and located at a distance of a tenth of a light year or so from the sun. Passing stars cause cometary objects from this halo to become deflected into orbits that take them to the inner regions of the solar system at the rate of about 1 or 2 every year. These deflected objects first show up as comets with long periods of revolution and a small fraction of them later become rounded up by the gravitational pull of the massive planet Jupiter into orbits of shorter and shorter periods. Comet Hyakutake is an example of a long-period comet, whereas Comet Halley has a period of approximately 76 years. There are comets of even shorter periods such as Comet Encke which has a period of 3.3 years. The Earth is well and truly entwined within these cometary orbits in such a way that material shed from comets must surely be reaching our planet quite plentifully. Such material is also injected onto the Earth by direct collisions with much smaller volatile comets. At a conservative estimate the amount of organic matter from comets actually arriving at the Earth could be reckoned on the order of thousands of metric tonnes per year.

Clumps of interplanetary dust particles of cometary origin have

[1]N.C. Wickramasinghe and F. Hoyle, Astrophysics and Space Science, 239, 121-123, 1996.

been collected in the stratosphere over many years using sticky paper flown aboard U2 aircraft. These so-called Brownlee particles (named after the investigator D.E. Brownlee) have consistently shown evidence of carbonaceous material, some of which might be exceedingly complex. Plate 12 (left panel) shows a micron-sized carbonaceous structure in a chondritic porous (CP) particle isolated and described by J.P. Bradley and his colleagues (*Science*, **223**, 56, 1984). These authors have interpreted this structure (which includes a magnetite domain) as being the result of 'heterogeneous catalysis'. However, comparing this with a microbial fossil found in the Gunflint cherts of N. Minnesota (right hand panel of Plate 12) we noted already in 1985 that a biological explanation (a partially degraded iron-oxidising bacterium) is perhaps more plausible (Hoyle et al, *Astrophysics and Space Science,* **113**, 20, 1985). In 1993 further studies by S.J. Clemett and his collaborators (*Science*, **263**, 721, 1993) of eight Brownlee particles, which were identified as cometary dust, revealed the presence of exceedingly complex organic molecules including aromatic and aliphatic hydrocarbons. This discovery was clearly a big step towards identifying cometary particles of the type shown in the left panel of Plate 12 as being biogenic.

A final and unequivocal proof of the biogenicity of cometary particles could, in our view, follow from NASA's Stardust Project. Preparations are underway for the launch of a spacecraft in 1999 to capture samples of dust from the short-period Comet Wild2 in the year 2004 for return to Earth in 2006. A rendezvous with the comet is expected to take place on 2 January 2004, some 98 days after perihelion passage. At this time the speed of the comet relative to the spacecraft will be about 6 kilometres per second, which although relatively low by normal standards will still destroy particles on impact on a hard surface. It is planned to catch the dust using a thick layer of aerogel, a silica-based spongy material that is 99% vacuum, in such a way that a soft landing will be achieved.

4

PRIMORDIAL SOUP AND RELATED MATTERS

4.1 INTRODUCTION

There is generally-agreed evidence that bacterial life was present on the Earth as long as 3.6×10^9 years ago, and there is good evidence of life going back even to the oldest-known rocks, 3.83×10^9 years ago. At the latter date, a banded-iron formation was deposited in what today is Western Greenland. To produce such a formation containing high concentrations of ferrous and ferric oxide (Fe_2O_3 and Fe_3O_4), there must have been a source of oxygen that was not atmospheric, in order to oxidize the normal iron oxide and iron sulphide (FeO and FeS) present in terrestrial rocks. Atmospheric oxygen can be excluded, it may be noted, because atmospheric oxidation of iron oxide and iron sulphide (FeO and FeS) leads to the deposition of red-coloured sandstones without high concentrations of ferrous oxide (Fe_2O_3) or ferric oxide (Fe_3O_4) being present in the sediments. The likely source of the needed oxygen was a by-product from such bacteria as *Pedomicrobium*, shown in Plate 13. Also shown in Plate 13 is an object discovered by Professor Hans Pflug within a sample of the Murchison meteorite (H.D. Pflug, in *Fundamental Studies and the Future of Science*, ed. C.Wickramasinghe, University College Cardiff Press, 1984). According to Pflug such objects are widely dispersed in the meteorite.

As we have already pointed out in Chapter 1, it has recently been found that the first indirect evidence of terrestrial microbial life shows up in the geological record about 3830 million years ago. This evidence is in the form of a subtle fluctuation in the ratio of carbon isotopes in carbonate deposits. Photosynthetic micro-organisms, which convert carbon dioxide in the atmosphere into biological material, have a slight preference for $^{12}CO_2$ compared to

$^{13}CO_2$. Thus biologically generated hydrocarbons and carbonates have a small deficit of ^{13}C compared with carbonates that are generated by non-biological means. Finding a deficit of ^{13}C relative to ^{12}C in ancient sediments can therefore be taken as evidence that photosynthetic life was present when the sediments were laid down. This is precisely what is found in the oldest terrestrial sediments, showing that photosynthetic microbial life already existed 3800 million years ago.

For those who believe that life originated on the Earth in a so-called organic soup, the slice of time available for the discovery of the genetic and enzymic complexity present in bacteria is reduced to a thin window. A truly immense stride in biochemical evolution would need to have been accomplished within the 800 million year time interval between the origin of the Earth 4.6 billion years ago and the date of the oldest presently exposed rocks. If still older rocks were discovered, this 800 million year interval would almost surely be squeezed from below, since it would be unlikely for the origin of life to have coincided exactly with the oldest rock sequence that just happens to be exposed at the present-day. And the 800 million year interval can certainly be squeezed from above, since the Earth at its formation was not in a condition to harbour life. The Earth had to acquire its volatile materials before there could be any possibility of life even existing, let alone originating here. The likely process whereby the Earth came to obtain its volatile materials has been described in earlier chapters, where it was seen to have arisen from the intricate dynamics of the condensation of the outer planets, especially of Uranus and Neptune, a process which has been estimated to have occupied the first 500 million years in the history of the planetary system.

In this context it is worth noting that the record of the early impact history of the Earth has only recently been unravelled by combining geological data with information derived from studying craters on the Moon. The data show that the Earth and the Moon were subject to frequent and violent impacts with cometary bodies up to about 3800 million years ago. We have already noted that the first signs of life appear very sharply at 3.83 billion years before the present, the time which also coincides with the moment when an initial huge surge of comet collisions had declined to a mere trickle. This remarkable coincidence, signifies in our view, that not only water and the gases of our atmosphere came from comets, but possibly the first microorganisms as well. For those who prefer the organic soup alternative it remains an incontrovertible fact that several 'squeezing factors' narrow the window for the hypothetical origin of life down to almost nothing at all.

Such unfavourable considerations have led members of the organic soup establishment to speak of 'instant life', implying that the astonishing biochemical complexity of life could arise all in a moment, just as it is supposed to have done by the creationists. That anything so absurd could be proposed at all is a consequence of the extent to which earlier absurdities of the organic soup establishment have been tolerated by a scientific world that has never been anxious, for the sake of its own peace of mind, to look into matters closely.

The amount of organic material present in the supposed terrestrial soup was miniscule compared with the amount which could have been present in the outer regions of the Solar System. If the hypothetical soup is taken to have contained as much organic material as there is currently in the Earth's biosphere the amount would have been $\sim 10^{18}$-10^{19} gm, a trivial amount compared to 10^{29}-10^{30} gm for the outer regions of the Solar System. Small-scale debris in the outer regions of the early Solar System could well have intercepted a considerable fraction of all sunlight, $\sim 4 \times 10^{33}$ erg s^{-1}, enormous compared to the amount of sunlight intercepted by the Earth, only $\sim 2 \times 10^{24}$ erg s^{-1}. Thus the availability of light for photosynthetic processes was also vastly larger in the outer regions, as was the availability of chemoautotrophic sources of energy and of radioactive energy. In relation to photosynthesis, it may be noted that maximum effciency occurs at light intensities significantly lower than the flux of sunlight at the Earth's distance from the Sun, and that photosynthesis can continue down to very low light intensities.

Commonsense would therefore dictate that a scientific establishment which seeks passionately to avoid the conclusion that life is a large-scale cosmic phenomenon, and which must therefore maintain that life originated in the Solar System, would do far better to put its trust in processes occurring in the outer regions, the regions of Uranus and Neptune and of the genesis of the comets, than to rely on a harsh rocky Earth initially even without water and without atmosphere. Our impression is that the 'life in the solar system' establishment would presently be glad to settle for this revised position, if only a way could be seen to avoid the issues still to be discussed in subsequent chapters, issues concerning the incidence of terrestrial diseases and of a driving evolutionary process from outside the Earth affecting the whole of terrestrial biology. These embarrassing concomitant matters are currently forcing the walking wounded of which this brigade is largely made-up to stay mewed in a siege condition, trying to maintain the old soup theory with their currently plaintive cry of 'instant life'. Otherwise sensible people think that our denunciations of the educational system are at best an amusing but cranky foible. This we believe to be wrong, for the educational system is responsible for producing a huge miasma of erroneous belief against which even the most sceptical and watchful of us are nine-tenths powerless to resist. Notorious examples arise from time to time, but one has to suspect that even the most flagrant of them is but a tip of a very large iceberg. Indeed one has to suspect that, so far as science is concerned, education has reduced mankind in the year 1997 to a condition not too far removed from outright insanity, a suspicion that is all too frequently confirmed by weekly science magazines, with their perpetually hysterical atmosphere of breathtaking progress, but which actually consist overwhelmingly of eyewash.

Those organic soup specialists who have claimed to produce organics of interest to biology from initially inorganic materials, by irradiating the latter with high-grade sources of energy such as ultraviolet light and accelerated particle streams, have certainly contributed a great deal of eyewash, brazen eyewash if such can be contemplated. Brazen from the outset, because some of the claimed inorganic materials were almost surely not inorganic at all. Almost certainly the methane (CH_4) used in so-called prebiotic experiments was of biological origin. So are more exotic sources of carbon such as ethylene, and so quite likely is the principal source of nitrogen, ammonia (NH₃). What should be considered so remarkable, a sane person can wonder, about producing substances of biological interest from what is already biological?

Doubtless the perpetrators of what, to put it plainly, has been a deceit would claim that methane (CH_4), could have been obtained by wholly inorganic means. But could it? The circumstance that nobody connected with prebiological experiments, and nobody apparently among the throng which has accepted the claims of such experiments, has bothered to discuss this question, or even we suspect to notice that it exists, shows just how far the whole scientific community persistently deludes itself. Without the dinning thud of the educational process we doubt that such an extreme measure of self-deception would be possible. Somewhere, among people left to think quietly without being avalanched by the weekly science magazines, the point would long since have been noticed, and the organic soup theory would long ago have been consigned to where it belongs.

It is now some years since one of the authors arrived at a helpful proposition where unsolved problems are concerned, namely that it is useless to follow popular opinion, because if solutions to unsolved problems lay where popular opinion holds them to be they would have been found already. To this we now add the useful dictum that wherever a long lasting confusion exists over the meaning of words something of significance lies waiting to be discovered. On this basis, the confusion over exactly what one means by 'inorganic' and 'organic' in chemistry deserves investigation.

Beginning with the Oxford Dictionary:

inorganic
Chem., of compounds not entering into composition of organized bodies; i. chemistry, that of mineral substances; not arising by natural growth; extraneous.

organic
Chem., of compounds existing as constituents of organized bodies, of hydrocarbons or their derivatives.

While these definitions touch our perceptions of what we mean by 'inorganic' and 'organic', they hardly approach the level of precision demanded in a scientific discussion.

Advanced texts usually take it for granted that we already know the meanings of the words. Thus *Advanced Inorganic Chemistry* (F.A.Cotton and G. Wilkinson, Interscience 1972) lives up to its title by plunging on the first page of the first chapter straight into group theory. As often in such situations an appeal to the *Encyclopeadia Britannica* yields as clear an exposition as probably can be obtained. The general article on Chemistry (15th Edition) has the following on its first page:

> *'Subdivisions of chemistry.* The field of chemistry encompasses the study of an uncounted and theoretically almost unlimited number of compounds. By the early 1970s there must have been more than 1,000,000 individuals working on chemical problems in independent, academic, industrial, and government laboratories throughout the world for a myriad of personal, social, economic, and political reasons. In systematizing chemical knowledge and activities by grouping together related compounds, related systems, related methods, and related goals, a number of subdivisions of chemistry have developed: These subdivisions provide the basis of organization of academic curricula and literature and of bringing together scientists who share common interest. During the first half of the 20th century, undergraduate curricula were almost exclusively organized into courses in inorganic, analytical, organic, and physical chemistry and biochemistry, which were usually studied in that order. This organization of subject material is still apparent in many college catalogues, but it is difficult to defend, and, accordingly, many attempts are being made to organize academic programmes along other lines.
>
> *Organic and inorganic chemistry.* Organic chemistry and inorganic chemistry are subdivisions based upon the elements present in the compounds. Organic chemistry is the chemistry of carbon compounds, which, of course, also contain elements other than carbon, such as hydrogen, oxygen, sulphur, nitrogen, phosphorus, and chlorine. Inorganic chemistry encompasses all substances that are not organic. The separation of the study of carbon compounds from the rest of chemistry is defensible on the basis of the sheer numbers of carbon compounds that are of great interest and that not only have been but are still being intensively studied. The structure of the carbon atom is unique among atoms and makes possible this great array of compounds, which are stable under atmospheric conditions on Earth but are also sufficiently reactive to make possible a great variety of chemical changes.'

The writer essentially reduces the division between inorganic and organic chemistry to a matter of arbitrary choice. One chooses 'organic' to be the chemistry of the compounds of carbon, and all other compounds are automatically 'inorganic' by definition. Without nuclear transmutations it would then be impossible to synthesize organic compounds from inorganic, for the evident reason that no inorganic material would contain carbon atoms. Substances like calcium carbonate ($CaCO_3$) would have been classed as 'organic', which of course would be against all precedent. This is not the way anybody thinks of carbon monoxide or carbon dioxide (CO, CO_2), or their associations in molecules like calcium carbonate.

Whereas the writer in the *Encyclopaedia Britannica* evidently defines the range of 'organic' substances too widely, the confinement of organics to hydrocarbons and their derivatives, as in the definition of the Oxford Dictionary, is too narrow. Amino acids are not hydrocarbons, nor is urea (H_2NCONH_2). The synthesis of urea from ammonium cyanate by Friedrich Woehler in 1828 is often mentioned at the very beginning of studies in organic chemistry, because it is supposed to have some deep significance. We have difficulty in understanding what the significance really was, although we have less difficulty in seeing what it was supposed to be. Urea is a biochemical, and what was claimed by people with a distaste for the old doctrine of vitalism was the synthesis of biomaterial from non-biological sources, a claim that was almost certainly false.

There is a great quantity of nitrogenous material in the soil, much of it in the form of ammonium salts of various kinds. While it would be difficult to prove with mathematical rigour that absolutely none of this material is of abiological origin, it is widely accepted that the bulk of nitrogenous material in the soil is the excretion products of denitrifying bacteria that operate to break down substances of biological origin. So if the ammonium cyanate in Woehler's synthesis came from a so-called natural deposit of some ammonium salt the chances are that the urea was obtained from material containing a biological product, in which case the logic of the anti-vitalist argument would be weakened to the point of nonsense.

It is clear from the example of the synthesis of urea that views on what constitutes 'inorganic' and 'organic' are by no means covered by unemotive definitions such as that attempted by the writer of the *Encyclopaedia* article. The relation of chemistry to biology is evidently involved, with chemistry seeking to claim a status independent of biology. In the sense that, given the elements in atomic form, they could be assembled under controlled conditions into small quantities of any organics one cared to specify, chemistry does have a status independent of biology. But in the sense that only very small quantities may be attainable in practice the issue requires further consideration.

4.2 A DEFINITION OF INORGANIC MATERIAL

The elements are not available in atomic form in quantity except in mass flows from stars and in the fraction of preplanetary material that became heated to high temperature in the solar nebula. By inorganic material we shall mean those compounds which form when such heated gas cools at low pressure (ca. 10^{-6} bar in preplanetary material) together with those more complex compounds that can arise abiologically when planets themselves are condensed. For example, magnesium oxide (MgO) and silica (SiO_2) condensing in cooling gas at low pressure can form magnesium silicate ($MgSiO_3$) when brought together at much higher pressures during planet formation. This example cannot be extended automatically to include all minerals, however, because quite likely bacteria were involved in the formation of some minerals. Each case must be considered according to the facts.

Iron and others of the transition elements condense as metals at approximately 1500K. Should the iron condensates become aggregated into fair sized blobs, the metal is preserved as cooling proceeds further, which is the likely route whereby the metallic core of the Earth orginated. As the temperature falls below 1000K, the thermodynamic balance for iron swings towards iron oxide (FeO) and iron sulphide (FeS), which happens for small particles and at the surfaces of large blobs. Other metals also emerge as oxides, aluminium oxide (Al_2O_3) for example. As the temperature falls below 1000K any remaining excess of oxygen goes to water (H_2O), while nitrogen emerges as molecular nitrogen (N_2) unless it can be argued that catalysis causes a reduction to ammonia in the presence of an excess of molecular hydrogen (H_2). Likewise carbon emerges as carbon monoxide (CO) or carbon dioxide (CO_2) unless catalysis produces a reduction to methane (CH_4) and other hydrocarbons. The free energy made available in the reaction nitrogen (N_2) + 3 hydrogen molecules ($3H_2$) → 2 ammonia molecules ($2NH_3$) is so little, however, only about 8kcal per mole of N_2, that the reaction does not go towards ammonia at a low pressure of $\sim 10^{-6}$ bar unless the temperature falls towards 100K. Because of the difficulty of maintaining suitable catalytic surfaces we take nitrogen to emerge in preplanetary material as N_2.

The free energy made available by the Fischer-Tropsch reaction with the basic form carbon monoxide (CO) + 3 hydrogen molecules ($3H_2$) → methane (CH_4) + water (H_2O) is about 35kcal per mole of CO, which permits the reaction to swing from left to right at a temperature of $\sim 500K$ or less. This is at the low pressure of the preplanetary gases. While such a temperature could in itself permit catalysis to occur, the reaction has proved difficult to oper-ate even under controlled industrial conditions, with great care over the preparation of catalysts being found essential. The Fischer-Tropsch reaction was actually used in Germany in the extremis of the 1939-45 war for the production of synthetic oil. Commonsense suggests that if the reaction is difficult at pressures very much higher than in the solar nebula, and with carefully prepared cata-lysts, it was hardly likely to have been effective in preplanetary material. Otherwise the price of oil would never have soared to ca.$30 per barrel.

Taking the production of methane (CH_4) and other hydrocar-bons to be ineffective for the reasons just stated, the present defin-ition of 'inorganic' compounds agrees with empirical practice to a remarkable degree, almost as if empirical practice had been based on considerations of star formation and planet formation. The notable exception is NH_3. We suspect that the the true inorganic form of nitrogen is N_2, not NH_3, which accords with the fact already noted above that the Earth possesses little, if any, inherent

store of the ammonium radical outside of that which has been produced by microorganisms. The conclusion also has a correspondence
with actual practice in industrial chemistry, where the production of
NH_3 by the Haber-Bosch process has long been regarded as a
watershed that leads to the production of organic materials that
would otherwise not be accessible commercially. The ammonium
situation is therefore critical and we shall discuss it in a little more
detail in a moment.

4.3 A CONJECTURE

The way things happen to be in nature, it is not possible by purely
chemical processes to pass in quantity from inorganic materials to
organic materials.

The restriction to chemical processes requires atoms to be conserved. This restriction is related to an exclusion from the argument
of biointellectual activities, as will be emphasized at the end of this
chapter.

THE HABER-BOSCH PROCESS

This process well illustrates the problems one encounters in trying
to find an example that disproves the above conjecture. The nitrogen used in the reaction nitrogen (N_2) + 3 hydrogen molecules
($3H_2$) → 2 ammonia molecules ($2NH_3$) is inorganic, but what of the
molecular hydrogen (H_2)? Early applications of the Haber process
obtained molecular hydrogen (H_2) by passing steam (H_2O) over hot
coke, which of course was of biological origin. Later applications
have obtained the H_2 from hydrocarbons of overtly biological
origin. But H_2 can also be obtained from the electrolysis of water
(H_2O), which is inorganic according to the above definition. So one
asks whence comes the electricity used for electrolysis? If from coal
or oil-fired power stations we are instantly back to biology. If from
nuclear reactors then processes outside chemistry have intruded. If
hydroelectric, the solar energy needed to lift water comes also from

nuclear processes in the Sun. In all cases we have considered, the route towards disproving the conjecture turns out to be similarly blocked. Exceptions can be devised but only it seems for very small quantities of material.

Extraterrestrially, free hydrogen must be considered as an inorganic material. Even so, conditions once again conspire against the production of ammonia (NH_3). Pressures in the preplanetary material are low. Temperatures in the atmospheres of planets such as Jupiter and Saturn are also low. Could one have the right kind of carefully-prepared catalyst, and could the catalyst be free-floating in a planetary atmosphere? In the absence of biology, conditions are not propitious. Microorganisms, on the other hand, possess exceedingly efficient catalysts, they are small enough to float in gases and liquids, and they can function down to remarkably low temperatures. Biology is a wholly dominant catalytic agent, and in its presence there can be no effective competition from abiological processes.

WHY THE CONJECTURE COULD BE A CRUCIAL MATTER OF PRINCIPLE

As cooling of the preplanetary gases proceeded below 1000K, the ensemble of molecules would go more and more out of thermodynamic equilibrium. This was because the Boltzmann factors affecting equilibrium increase in importance as the temperature falls. At high temperatures thermodynamic equilibrium favours there being as many gas molecules as possible, which leads to inorganics like carbon monoxide (CO) and molecular nitrogen (N_2) being dominant. At sufficiently low temperatures, however, the positive free energy values obtainable in the formation of methane (CH_4) and ammonia (NH_3) favour the reduction of carbon monoxide (CO) and molecular nitrogen (N_2) in the presence of an excess of H_2. But according to the point of view developed above, the reactions nitrogen plus hydrogen → ammonia ($N_2 + 3H_2 → 2NH_3$), carbon monoxide plus hydrogen → methane plus water ($CO + 3H_2 → CH_4 + H_2O$), do not proceed in a purely inorganic situation. When the

temperature has fallen low enough for them to be thermodynami-
cally preferred, their rates are then too slow unless highly specific
catalytic surfaces are present, which seems unlikely to be maintained without the intervention of biointellectual processes. So cooling preplanetary gas at low pressure goes out of chemical balance;
and it does so for some of its commonest molecules. According to
our point of view it is this situation, applicable everywhere not just
in the Solar System, which creates the primary niche for biology.
Biology is 'nature's way' of moving much closer to thermodynamic
equilibrium than would otherwise be possible. Stated the opposite
way, if thermodynamic equilibrium could be reached abiologically,
if inorganics could go to organics abiologically, if our conjecture
were substantially untrue, biology would be short-circuited by the
inorganic world and could not exist.

The physical *raison d'etre* for biology is thus seen to be the widespread deviations from thermodynamic equilibrium that would
otherwise exist on a universal scale, deviations which occur whenever hot material cools at protostellar pressures. Even if there were
not many facts which show biology to be a universal phenomenon,
the notion that the exquisitely complex enzymic systems of biology
exist on the Earth alone, in order to cope with only a local depar-
ture from thermodynamic equilibrium, could be seen to be improbable if not indeed absurd.

Anaerobic microorganisms are likely to be found as chemo-
autotrophs living everywhere in deviations from thermodynamic
equilibrium, provided the twenty or so chemical elements necessary
for life are available in the localities in question, of which there are
a number at the present day in our own Solar System as we saw in
Chapter 1. Of such possibilities, only the Moon (where the requisite
elements are not all available) and Mars have so far been explored
directly, the latter only cursorily. Even so, many who have examined
the details of the Viking missions to Mars have now concluded that
the missions were inconclusive and further explorations are planned
to take place early in the new millennium.

All this has been subject to the chemical restriction that atoms
be conserved. Although biology cannot it seems be by-passed in its
control over thermodynamics in cooled material when atoms are

conserved, biology can indeed be by-passed if this chemical condition is broken. Given nuclear power, H_2 becomes available terrestrially through electrolysis, when an appreciable route from the inorganic to the organic at last exists. Even so be it noted, biology has not been backward in incorporating the fruits of nuclear processes into its repertoire. More than 10^{17} grams of biomass are produced on the Earth through photosynthesis annually. It will be a long time before human technology produces 10^{11} tons of organics per year. Without photosynthesis i.e. without nuclear input, biology is confined to comparatively low-grade chains for energy production. With photosynthesis, biology makes use of higher grade forms of respiration as for instance glycolysis. It seems clear that biology has used the lower forms to step-up to the higher forms.

The advent of nuclear fission has had a profound effect on our human social systems. Perhaps the fear of people of nuclear activities in general has something more in it than a straightforward fear of the explosive violence of nuclear weapons. In by-passing biology in a deep thermodynamic sense, nuclear power could in principle provide a mode of organized existence that was independent of biology. One can imagine a nuclear-powered robot society, computer-driven, existing independently of biology. Not tomorrow as in science-fiction. But why not in 50,000 years? With the discovery of nuclear power a crucial dividing line has been crossed. It is conceivable this crossing is vaguely perceived by people at large, and the perception of it may lie at the root of present unease, which may even be preprogrammed within us, as great abilities in mathematics and music appear to be preprogrammed. Preprogrammed by biology as a defence against a dangerous new rival.

The superpower confrontation until the mid-eighties proceeded along what appeared to be laid-out guidelines, with two superpowers moving step-by-step rather as a couple executes a dance defined by chalk marks on a ballroom floor. The music grows louder, the beat more insistent, until the two are impelled together, whether they want to be so or not, to find themselves in a particle-antiparticle collision. Or like driving in a steadily thickening fog. Up to a point the driver remains in control. Then quite suddenly control becomes very difficult, and unless one somehow gets off the road a crash becomes inevitable. Mathematicians and physicists understand stability phenomena like this very well, and have no difficulty in perceiving that the progressive shortening of the time available for a response to be made to a first-strike nuclear attack is analogous to a thickening fog on the road. Perhaps the vehicles can be slowed and edged off the road before the fog becomes too thick. If not, a failure in the educational system to understand the broad principles of the interrelation of biology with physics and chemistry will in our view have played no small role in provoking the eventual disaster.

The last part of the above discussion may prove fanciful, and we hope it will. The concept of an entirely robot-controlled society may well also be fanciful. We have included it here so as not to exaggerate the strength of our chemical conjecture. If such a possibility were excluded, our conjecture could be upgraded, because devices such as nuclear reactors and turbines for hydroelectric power, being man-made, would then become biointellectual features that were excluded unnecessarily in the formulation of the conjecture. Including biointellectual activities alongside biochemical processes, the restriction on atoms being conserved may be abandoned, and we would then assert the strong proposition that it is impossible to pass in quantity from inorganic materials to organic materials except through the intervention of biology.

5

BEGINNINGS OF BIOLOGY IN COMETS

5.1 INTRODUCTORY REMARKS
Uranium-eaters and chain-reactions

However often one learns to accept the amazing ways in which biological systems make use, not only of their gross environment but of subtle aspects of physics and chemistry, one never becomes quite inured to new surprises. It was so with us on the day we learned in private communication that some species of bacteria can precipitate uranium salts from very weak solutions, 'they are practically uranium-eaters' our acquaintance told us. A possible answer to an unresolved conundrum over the Oklo reactor occurred to us thereupon to which we shall return in a moment.

Once again it was a surprise when we learned from Drs. R.B. Hoover and M.J. Hoover that some species of diatoms are able to concentrate other potentially fissile elements:

'. . . it is established that many species are capable of thriving in environments containing extremely high concentrations of unusually lethal radioisotopes such as americium, plutonium, strontium, etc. Diatoms thrive in highly radioactive ponds, including the U-pond and the Z-trench at the Hanford facility, with the latter containing over 8kg of various radioisotopes of plutonium. Not only do diatoms live in this environment, but they seem to have a remarkable affinity for plutonium (c.f. R.M. Emery, D.C. Klopfer and W.C. Weimer, 1974, in *Report prepared for the U.S. Atomic Energy Commission under Contract AT(45-1)*: 1830, BNHL-1867 p.44). The algae of these ponds, of which diatoms are by far the dominant form concentrate [241]Am three millionfold, and certain isotopes of plutonium are accumulated to 400 million times the concentration in the surrounding water. The plant life in these radioactive ponds contains more than 95% of the total plutonium burden. Diatoms and *Potamogeton* alone contain more than 99% of this plutonium. In such an environment, diatoms grow in great abundance while continuously subjected to high levels of x-rays, gamma rays, alpha and beta particles . . .

Deaths from leukaemia tend to show a peculiar very local clus-
tering effect. We recall the example of a remote valley in New
Zealand where over a time scale of a few years there were about 10
such deaths, a valley where there was no nuclear reactor. A cluster
of six similar cases later came to public notice in the village of
Seascale, close by the Sellafield nuclear reactors of West Cumbria.
Somewhat naturally, the media have attributed the latter unfortunate deaths to the presence of the reactors, and as an outcome of
media pressure the Ministry of the Environment of the British
Government was led to set up a committee of enquiry into the matter. Members of the committee knew perfectly well from fully
attested statistics that neither the natural radioactive background
nor the slight increment in the background caused by the reactors
could explain the facts (except as a truly monstrous statistical fluctuation) and they also knew, which their critics apparently did not,
of the existence of similar clusterings elsewhere, as in New Zealand
where there had been no nuclear reactor. So the committee simply,
and somewhat innocently, reported that the Sellafield reactors could
not have been responsible for the six leukaemia deaths. Naturally
London journalists writing for British weekly science magazines,
who frequently meet together at various gatherings and perhaps
over a pub lunch or two, and so tend to be of a single mind on such
issues, had a field day over it. What the journalists could see with
startling clarity, just like everybody else, was that while leukaemia
cases might occur in small clusters it was apparently most peculiar
that one such cluster should be found sitting almost on top of a considerable complex of reactors, where for one reason or another the
management had not over past years been able to avoid the escape
of small quantities of radionuclides into the local environment.

Diatoms do not rate highly in the educational system, and
topics with low priorities receive scant attention in both books and
in lecture rooms. Quite likely therefore, the facts concerning
diatoms reported to us by Drs. R.B. and M.J. Hoover are unknown
to the umpires of the weekly science magazines, to the media, and
possibly to offficialdom decked in its magisterial robes. Otherwise
there would surely have been a rush to examine the water supplies
at Seascale, not with a view to its dissolved contents, but with

respect to microorganisms suspended within it. Especially as the nearest Lakeland valley to this part of the Cumbrian coast is Eskdale, where granite rocks containing a high level of uranium (and so of the decay products of uranium, some alpha-active) outcrop the surface. If cultures in water pipes concentrate such products in the manner described above, with local populations imbibing the microorganisms, perhaps with a further concentration occurring in the human body itself, the facts would become intelligible. One reason for the siting of the Sellafield reactors was related to the water supply from Eskdale and Ennerdale, it is ironic to notice, and this might be the connection the media have been seeking, an innocent connection that would not be much to their liking if it turned out to be true. Similar conditions obtaining elsewhere would of course produce the same effect, regardless of whether there were nuclear reactors in the districts in question.

Concerning the Oklo reactor, Dr. S.A. Durrani of the University of Birmingham wrote as follows:

'Nature, it would seem, had anticipated man by something like 1,800 million years in bringing about the first self-sustained nuclear chain reaction on the Earth. And, contrary to common belief, it was not in the squash court of the University of Chicago in December 1942, but in the wilds of what is today the Republic of Gabon at a place called Oklo that this fantastic phenomenon took place.

'The history of the discovery of the phenomenon, as it unfolded during the symposium, is briefly as follows. In June 1972 a team working under the direction of Dr. H.V. Bouzigues at the CEA service laboratory at Pierrelatte in France noticed a marked anomaly in the abundance of the uranium-235 isotope (0.7171 $\pm$ 0.0010 in atomic per cent instead of the normal 0.7202 $\pm$ 0.0006) during the certification of a secondary standard of UF_6 by the gas diffusion method. Later, much larger depletions of this isotope were discovered (down to 0.621%, and eventually to 0.296% U-235) in uranium samples from this source, which was traced back to the Oklo deposit. First positive proof of the hypothesis that a natural chain reaction was responsible for the depletion of the fissile component was furnished by Mme. M. Neuilly and co-workers of CEA through the measurement of the ratios of fission-product rare earths detected in the ore by the spark source mass spectrometry technique. Two simultaneous submissions by the above two groups on September 25, 1972, to the Proceedings of the Academy of Sciences, Paris, announced the discovery and the proposed explanation of this remarkable phenomenon. It was pointed out that at the time of the reaction the natural abundance of the relatively fast-decaying ^{235}U isotope was more than 3%. This natural 'enrichment', helped by the moderation of the fission neutrons by the water content of the soil which enhanced their fission efficiency, and possibly by the relative absence of neutron-absorbing elements in the surroundings, allowed a nuclear chain reaction to develop. It is perhaps worth mentioning that such a natural chain reaction had already been predicted, on theoretical grounds, by several scientists, notably by P.K. Kuroda as early as 1956. The scientific secretary of the symposium, Dr. R. Naudet of CEN, Saclay, has since late in 1972 been leading the 'Franceville Project' established by the French CEA to investigate the phenomenon, and has done a great deal to promote its study internationally . . .'

The first announcement from the CEA laboratory provoked scepticism among nuclear physicists, because of the point alluded to briefly in the above quotation, the need for an absence of 'neutron-absorbing elements in the surroundings'. Very little in the way of elements such as cadmium or gadolinium would have poisoned the reactor, and the difficulty was to see how under aqueous conditions all such neutron poisons could have been conveniently absent. What happened subsequently was that French physicists gathered sufficient evidence concerning the presence of fission products at the site of the reactor to convince the sceptics. But without the problem of neutron poisons being cleared up satisfactorily.

The statement that some bacteria are 'practically uranium eaters' suggested both a possible cause of the Oklo phenomenon and a resolution of the neutron poison problem. Imagine bacteria in comparatively still water precipitating around themselves a high density coating of increasing thickness of some uramum salt, uranium oxide most likely, rather as bacteria precipitate calcarious material to produce stromatolites. The increasing coating would eventually cause the bacteria to sink to the bottom of the lake or pool in which they had been suspended. In the floor of the pool, suppose there to have been a bowl where more and more uranium-coated bacteria accumulated. A stage would be reached at which the growing colony went critical in the manner of a simple boiling water reactor using enriched uranium, the 'enrichment' for ^{235}U being of the order of three per cent at the epoch of the Oklo reactor.

The violent motion associated with boiling could scatter the bacteria in a timescale less than the interval of about 30 seconds required for the appearance of the main complement of delayed neutrons, thus maintaining stability should such a system threaten to become seriously supercritical.

All living systems produce a great measure of chemical segregation, accepting some elements and rigorously rejecting others. We have never heard of the elements gadolinium or cadmium being present in living organisms for example. In this way one could elegantly understand the absence of neutron poisons from a biological reactor, thereby overcoming the previously mentioned difficulty which at first sight had erroneously seemed to obviate the French discovery.

If a so-called 'natural' reactor could arise 1800 million years ago, when the enrichment of ^{235}U was about 3 per cent, bioreactors could arise almost trivially one might suppose in the early days of the Solar System when the enrichment was about 30 per cent. This likelihood raises the possibility of an escape for life from the present-day straightjacket of temperature, the slim zone here on the Earth between being boiled alive (Venus) and being frozen solid (Mars). With a controlled heat source inside an adequately insulated body, life on the outside of the Solar System in its early history could have adjusted temperature conditions to suit itself.

The most probable sites were planetesimals of various sizes, from a lunar scale down to a cometary scale, with liquid interiors generally at temperatures of around 300K, surrounded by surface shells of frozen material having low heat conductivity. Superinsulators with porous structures have coefficients of heat conductivity $\sim 10^{-4}$ watt cm^{-1} K^{-1} (c.f. J.E. Parrott and A.D. Stukes, *Thermal Conductivity of Solids*, Pion Ltd., 1975, 143), a value that will be used in the following discussion.

The heat-release process was the one already described above, determined by the precipitation of the potentially fissile elements U, Th, by microorganisms that subsequently sank towards the centres of the planetesimals where they contributed together to produce a critical reactor which stabilized itself by generating convective motions that mostly prevented the central concentration of fissile material from attaining a runaway supercritical condition (although a recollection of Oort's exploding planet flickers in one's mind at this point).

To estimate the potential amount of fissile material, it seems reasonable to suppose that breeding of ^{232}Th and ^{238}U to ^{233}U and ^{239}Pu respectively could occur in a large measure – from the point of view of reactor technology this should have been 'easy' at a time when the ^{235}U enrichment was so high. Solar abundance tables by numbers of atoms give (U + Th)/(C + N + O) $\cong 6 \times 10^{-9}$, which is a ratio by mass $\sim 10^{-7}$. This estimate rests on the amounts of U and Th actually found in meteorites, however, which raises the possibility that what has been measured for meteorites are low values subsequent to appreciable denudation by bacterial action. Calculations based on the so-called r-process for the primordial genesis of U, Th have run an order of magnitude higher than the measurements, and here we may well have the reason for this discrepancy – the meteoritic values are not primordial, thereby destroying a hallowed assumption of meteoritic chemists, which by a refusal to question it has attained the status of a religious dogma. If we take an intermediate position, with (U + Th)/(C + N + 0) $= 3 \times 10^{-7}$ by mass, we shall not be far wrong.

The output of energy from the total fission of U + Th is $\sim 10^{18}$ erg g^{-1}. Hence with most of the mass of material being C, N, O, on the outside of the Solar System, the fission energy yield per gram from its content of U + Th would be $\sim 10^{18}$. $3 \times 10^{-7} \cong 3 \times 10^{11}$ erg. With 10^{29}-10^{30} gm of C, N, O, the total energy available was therefore $\sim 10^{41}$ erg.

With the material having a density ~ 1 gm cm^{-3}, the mass of a body of radius R was $\sim 4\pi R^3/3$, and the total energy available for release inside the body $\sim 3 \times 10^{11} \times 4\pi R^3/3$ erg, with R in cm. Suppose such a body to have a surface shell of thickness 1 km through which the temperature fell from 300K on the inside to $\sim$ 100K at the outer surface. With a heat conductivity of 10^{-4} watt cm^{-1} K^{-1} the heat loss through the shell would be $10^3 \times$ $(4\pi R^2)$ $(200/10^5)$ erg s^{-1}, with R again in cm. The heat availability is sufficient to make good this loss for a time T seconds given by

$$10^3 (4\pi R^2) (200/10^5) \; T = 10^{11} \times 4\pi R^3. (1)$$

Or with T in years and R now in kilometres,

$$T \cong 2 \times 10^8 \; R \; \text{years} \hspace{4cm} (2)$$

There would evidently be no difficulty for a body of lunar size, R >1000 km, maintaining a liquid condition in its interior, and some comets might have been able to do so over at least the first 500 million years in the history of the Solar System. Excess energy output would simply lead to a thinner surface shell, while a reduction of output would thicken the shell, in effect with the shell thickness adjusting itself to the reactor output.

This solves the problem for the existence of chemoautotrophic biological systems under anaerobic conditions. Or, in other words, we have a possible *modus operandi* for biological systems that build their own nutritive substances by anaerobic chemosynthesis. If we reckon 3×10^{10} erg gm^{-1} as the average chemical energy available for chemoautotrophy, then the total for the whole outer Solar System is $\sim 10^{40}$ erg, about an order of magnitude less than the radioactive energy, but still a very large amount. It is here that the considerations of the previous chapter become relevant, that there should be no way to unlock this great store of energy except through biology.

Unlocking the store of chemical energy degrades the material in a thermodynamic sense, which consideration raises a further critical question: Could there be any means of achieving photosynthesis and so avoiding the progressive degeneration due to chemoautotrophy? Very readily, provided fibre optics existed to channel light through the cold surface shell to the reservoir of warm liquid below. A little thought shows that such a possibility is not as fantastic as it might appear at first sight. Since biology has produced eyes with their ability to function over an exceedingly large light intensity range, eyes with sophisticated chromatic and spherical aberration corrections included, and with such acuity of focus that a bird can distinguish small scraps of food from unwanted debris at distances of several hundred metres, fibre optics should not have been any great obstacle. From a physical point of view, such a requirement amounts to combining translucence with low heat conductivity, and also with a high opacity in the infrared, conditions that together would permit the surface shell of material to act as a powerful greenhouse, thus easing the load on the internal heat production considered above, and extending the estimate (2) for the length of time T over which biological activity could continue.

A greenhouse effect could reduce very greatly the thickness of the required outer shell, making the penetration of visible light much less of a problem. Light penetrates typically about 10 metres through many translucent materials, for example, through water-ice. For a fall of 200K through a 10 metre thickness of material with heat conductivity 10^{-4} watt $cm^{-1}K^{-1}$ the heat flux is 2×10^2 erg cm^{-2} s^{-1}, which equals the flux of sunlight at a heliocentric distance of ~80 AU. This is for sunlight at normal incidence. On a rotating body the generally oblique incidence of sunlight (and no sunlight at all on the dark side) reduces the average flux by 4, so decreasing the corresponding calculated heliocentric distance by 2, from ~80 AU to ~40 AU, i.e. to the outskirts of the present-day distribution of planets. Hence there seems no reason in principle why a vast biological ensemble should not have persisted on the outside of the Solar System over an extended period of several hundred million years, and why it should not have done so in an ongoing replicative state. There seems no reason also why life forms on the Earth, especially among invertebrates, should not have been derived directly from this former condition, assuming a feasible form of transportation being available from the outer regions of the Solar System to the Earth. Comets perturbed by stellar encounters into orbits with perihelion distances q < AU are the obvious candidates for such a means of transportation. There appears to be no reason either why bubbles of gas should not become established as vacuoles within the objects, so permitting subaerial biological forms to arise. If the present-day complexities of life can arise by evolution in a biosphere of only ~10^{18}-10^{19} gm, the possibilities for a supersystem with mass ~10^{29}-10^{30} gm would almost surely be immense, especially as collisional interchanges which must have taken place from time to time among the many objects would have permitted evolutionary steps to be widely-shared among them.

On this view, comets are relics of a former large-scale biological environment existing in the outer regions of the Solar System. The total mass of the relics, say 10^{11} comets of individual masses $\sim 10^{18}$ gm, again enormously exceeds the terrestrial biosphere. The total cometary storage of biomaterial could be as high as $\sim 10^{29}$ gm, and it would be surprising if this large quantity of material had not dominated conditions at the terrestrial surface throughout the history of the Earth. We tend to think the opposite simply because the total mass of the Earth, $\sim 6 \times 10^{27}$ gm, is much greater than the mass of an individual comet. But the total mass of the Earth is an irrelevancy here. It is the mass of the terrestrial biosphere that in the present discussion really counts, and the biosphere matches only a single comet out of the 10^{10} comets which must have passed through the inner regions of the Solar System during the history of the Earth. The weighting factor in favour of comets controlling the evolutionary situation is evidently enormous.

We end this chapter by asking what happens should the nuclear engine inside a comet finally give out? With the internal heat source gone, and yet with heat losses continuing at the surface, the comet must eventually become cold and frozen throughout its interior. If water is an appreciable constituent, a liquid interior inside a solid shell could not freeze without dramatic events occurring simply because of the volume expansion that water undergoes on freezing. First, as the engine gave out convection currents stirring the liquid would cease, and all solid particles hitherto suspended in the liquid would fall gently towards the centre. Among the particles could be small silicate grains together with other refractories as well as microorganisms. Hence an aggregate of particles would be deposited by sedimentation, as the carbonaceous meteorites have been formed by sedimentation, and with an admixture of microorganisms as Hans Pflug finds to be present in these meteorites. The sedimentation would proceed higgledy-piggledy, just as the small particles happen to settle out of the now unstirred liquid.

Freezing goes progressively from the outside inwards. Water

immediately inside the outer shell cannot freeze without space being created for it to squeeze into. However, unlike water freezing downwards in a lake, which can simply lift the surface skin of ice bodily in order to create the needed space, water inside a closed frozen shell cannot lift the shell without cracking it into two or more parts. This requires a pressure of the order of the tensile strength of hard-frozen ice to develop throughout the liquid interior, $\sim 3 \times 10^7$ dyne cm^{-2}, a pressure which then acts compressively on the central concentration of small particle sediments, just as the carbonaceous meteorites were acted on compressively by a pressure of this order.

If freezing were a discrete one-step affair, an entire cracking of the outer shell into two parts might happen, but with the freezing process occurring continuously, a steady pressure $\sim 3 \times 10^7$ dyne cm^{-2} would be maintained against the inner surface of the shell. and within small cracks as they opened up, probably in many places throughout the shell. The lowest density components of the liquid would be squeezed up into the cracks, and likely enough would eventually emerge at the outer surface of the shell. In such a continuous multicracking process the needed extra space to provide for the expansion of the water would be found through geyser-like spurts of liquid, up through newly-opened cracks, with the liquid welling out and eventually freezing on top of the shell, the needed space being thus found on the outside of the comet. Since there are many organic liquids with densities less than water it would be these that would pour out of the freezing comet in preference to water, so explaining a generally observed situation in comets, where highly volatile organic materials are located close to their outer surfaces. In this way some of observed activity in comets at large perihelion distances (eg. Comet Hale-Bopp, and Comet Halley at >6AU) could be explained.

The final picture to emerge of a frozen comet is not of a single fused solid ball, but of an exceedingly complex multicracked affair, with the whole comet internally stressed at $\sim 3 \times 10^7$ dyne cm^{-2}. The situation is analogous to a mass of coiled springs, all ready to go off at a touch, which happens whenever evaporation due to sunlight weakens particular holding points in the structure. Or like a group of drunks leaning on each other – take one out and repercussions are felt throughout the whole party. On this picture a comet would not be exactly the most restful place one might visit. Comets which approach close to the Sun often lose fair-sized chunks of themselves, which separate apart quite gently at speeds of no more than one or two metres per second. If comets were homogeneous solids, the tensile strength of the material would have to be less than 10^5 dyne cm^{-2} to permit this phenomenon (Z. Sekanina in *Comets*, ed. L.L. Wilkening, University of Arizona Press, p.251). Since no well-frozen solid material has a tensile strength remotely as low as this, we can conclude that either a comet is a multicracked ensemble, with bits of itself only very lightly attached to other bits, or the interior material is still liquid.

With R = 5 km, equation (2) gives T = 10^9 years. This estimate for the time scale over which a cometary nuclear engine could maintain a liquid condition in the interior is so close to the ages of comets that we might reasonably argue both ways, with smaller comets having undergone freezing, and with larger ones still maintaining liquid interiors, and perhaps still maintaining something of their original biological activity. We are tempted to associate P/Schwassmann-Wachmann I with this condition. The sporadic outbreaks of this comet can then be understood in terms of an accumulation of biochemically-produced gas within the interior, pockets of which break from time to time to the surface expelling visible clouds of gas and particles, perhaps in a similar fashion to the generation of dust storms on Mars. A similar type of behaviour has also been observed more recently in the case of several other comets.

Comet Halley was found to produce prodigious output of particles which continued even when it had retreated beyond the orbit of Jupiter. A dust halo was observed when the comet was half way between the orbits of Saturn and Uranus. Millions of tonnes of organic dust continued to emerge out of the nucleus of Halley. Biological activity building up sub-surface high pressure pockets that periodically explode would seem to provide a natural explanation of all these facts. A very similar pattern of behaviour has been repeated in recent years for other comets, notably in the case of the comet Hale-Bopp which had already, in August-October 1995, developed an oblong coma of dust at the radius of Jupiter's orbit. (See Plate 14).

A class of comet that has recently come to the fore is that of the 'giant comet', and such comets with typical radii of 30-200km could be particularly relevant to the preceding discussion. According to equation (1) such giant comets would be able to retain liquid interiors throughout much of the history of the solar system. It is also possible that such giant comets may be associated with a primordial reservoir of comets, as first suggested by G.P. Kuiper in the 1950's, that would be longer-lived than the Oort cloud to which we have already referred. The idea of giant comets, originally proposed by Victor Clube and Bill Napier, remained only a conjecture until the discovery of Chiron in 1977. At first it seemed that Chiron was just another asteroid in an orbit that lay mostly between the orbits of Saturn and Uranus. In 1989, however, Chiron brightened so greatly and even developed an extended (5 arc second) coma that it subsequently came to be regarded as a giant comet. More than two dozen similar objects have been discovered to date, and they are mostly present in the outer regions of the solar system, near Neptune and Pluto.

Comet Hale-Bopp

The giant comet that has made headlines in recent months is the Comet Hale-Bopp to which we already briefly referred (Plate 15). This comet which has a diameter of some 40km (25 miles) falls towards the lower end of the size range proposed for giant comets. It cruised into the inner solar system in a highly elongated orbit, the plane of which was almost at right angles to the plane of the Earth's orbit. (See Plate 16). The orbital period as it came in from the depths of the Solar System was 4200 years, but its passage past Jupiter at a distance of 65 million miles resulted in its period being reduced to 2380 years, which would make the comet reappear in our skies in the year 4377AD[1]. Comet Hale-Bopp brightened with a steeply rising light curve as it came in to perihelion on April 1.

---

[1]When comets in their orbits come sufficiently close to Jupiter they can have their orbital periods shortened due to the gravitational pull of this planet.

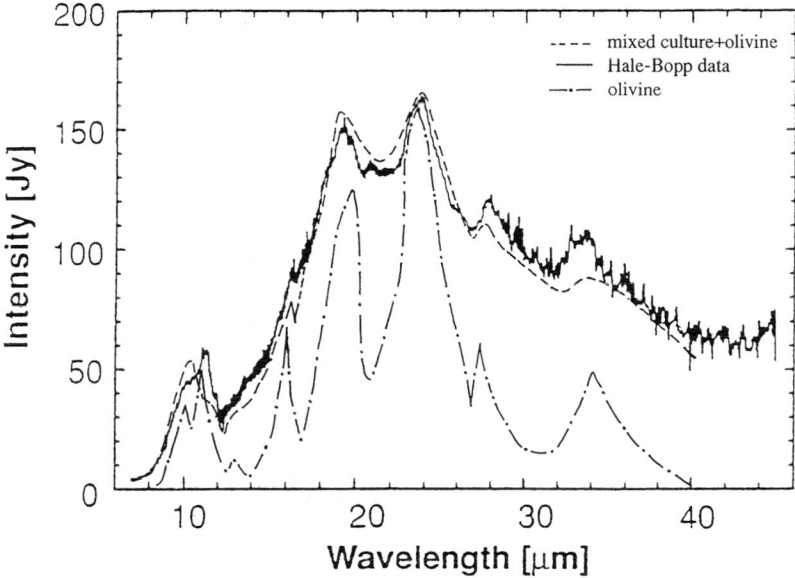

Fig. 5.1: The spectrum of Comet Hale-Bopp at 2.9AU from the Sun (jagged curve) compared with the predictions for a mixture of a microbial system (temperature 200 degrees Kelvin) together with 10% by mass of inorganic olivine dust (temperature 175 degrees Kelvin). The dashed curve is for pure olivine grains heated to 200 degrees Kelvin.

With a tail extending over 10-30 degrees, implying a dust column of several millions of miles in length, it appeared in the sky night after night for several weeks in March and April 1997. Comet Hale-Bopp passed within 194 million kilometres (122 million miles) of the Earth on March 23, 1997, and reached its closest to the Sun, about 138 million kilometres (85 million miles) on April 1st. The comet crossed the ecliptic plane (plane of the planetary orbits) at a distance of about 8 million miles from the Earth's orbit. If a sufficiently wide meteor stream followed the comet in its orbit, the Earth would be expected to intercept this stream towards the end of 1997.

At the time of going to press various aspects of this comet are being hotly pursued by astronomers and space scientists the world over. Many types of molecules, including organic molecules, have been discovered in the comet's coma; jets of dust have been found to emanate from the nucleus, which is rotating with a period of close to 11 hours. J. Crovisier and his colleagues have used the

Infrared Space Observatory (ISO)[1] to observe the comet when it was at a distance of 2.9 Astronomical Units from the Sun. Their observed spectrum is displayed as the jagged curve of Fig. 5.1. The dashed curve shows our calculation for a model involving approximately 90% by mass of a general bioculture and 10% in the form of olivine dust.[2] The dot-dash curve is the calculation for olivine (crystalline magnesium silicate). Although inorganic olivine is able to explain the positions of the peaks in this Figure, it can be seen that olivine alone is hopelessly inadequate to explain the data. A dominant contribution from a microbial system (including diatoms, which are algae) seems to be required. Just as with the explorations of Comet Halley in 1986, there could be many surprises that lie ahead, and many a cometary model might be overturned.

Comet Shoemaker-Levy

Another recent comet that is relevant to the ideas discussed in this chapter is Comet Shoemaker Levy 9. This comet, with an initial diameter about 9 kilometres, was fractured into some 20 separate pieces in July 1992 when it passed close by Jupiter, and all the fragments subsequently plunged headlong into Jupiter. Studies of the impact sites showed evidence of submicron-sized particles, possibly of organic composition, but because Jovian gases could also have recondensed, the situation regarding the original composition of SL-9 dust is not entirely clear. However, the break-up event itself indicates that the comet was not a solid piece of ice, but a complex and fragile assemblage of 'icebergs' including volatile organics exactly as we have discussed earlier. Plate 17 shows a picture of the many pieces of SL-9 before their final plunge into Jupiter. The event depicted in this Plate has caused scientists to take seriously the possibility of cometary impacts on the Earth, a matter to which we shall return in the Epilogue.

[1] J. Crosvier et al, Science, 275, 1904, 1997.
[2] C. Wickramsinghe and F. Hoyle, Natural Science, Vol 1, 1997.
(Internet – http://naturalscience.comm.ms/ns.home.html)

6

MICROBIAL INVASIONS FROM SPACE?

6.1. EVIDENCE FROM MEDICAL ANNALS
Common cold

We began our enquiry into the history of diseases in a curious way. We were unable to recall any mention by Shakespeare of the common cold. *The Oxford Dictionary of Quotations* has 67 pages quoted from Shakespeare, but nothing related to the common cold, although later authors with only a page or two in the *Dictionary* do make clear references to 'cold' as an affliction. Shakespeare's use of 'cold' always relates either to death or to the physical act of being cold. We could think of instances where 'fever' was used, but it is characteristic of the common cold that a clearly noticeable elevation of temperature is comparatively rare.

Yet the common cold is not a new visitor to our planet, since there is the following description of it in the Hippocratic writings:

'In the first place, those of us who suffer from cold in the head, with discharge from the nostrils, generally find this discharge more acrid than that which previously formed there and daily passed from the nostrils; it makes the nose swell, and inflames it to an extremely fiery heat, as is shown if you put your hand upon it. And if the disease be present for an unusually long time, the part actually becomes ulcerated, although it is without flesh and hard. But in some way the heat of the nostril ceases, not when the discharge takes place and the inflammation is present, but when the running becomes thicker and less acrid, being matured and more mixed than it was before, then it is that the heat finally ceases.'

Evidently there were sufferers from the common cold in classical Greece, just as there are today. Nowadays more working days are lost from colds than from any other cause, a situation that possibly did not obtain in Shakespeare's time. It is true that Marion's trouble in the following passage from *Love's Labour's Lost* sounds rather like a cold, although there are other possible interpretations:

> 'While greasy Joan doth keel the pot.
> When all around the wind doth blow,
> And coughing drowns the parson's saw;
> And birds sit brooding in the snow,
> And Marion's nose looks red and raw,
> When roasted crabs hiss in the bowl.'

Yet if indeed colds were then widespread, it is surprising that Shakespeare did not make use of them in his comic scenes.

Smallpox

The literary evidence for smallpox being a periodic visitor to our planet is stronger than for the common cold. A well-known medical text that we consulted at an early stage contained the remarkable statement:

> 'In the sixth century smallpox prevailed and subsequently, at the time of the Crusaders, became widespread.'

One could wonder how a disease as infectious as smallpox could have 'prevailed' in the sixth century and yet not have become widespread at that time. After telling the reader that the disease 'spreads like fire in dry grass', the text went on to say:

> 'The disease smoulders here and there and, when conditions are favourable, becomes epidemic.'

One can also wonder how it came about that something which spreads like fire in dry grass managed to smoulder.

The etymology of the Latin word *variola*, meaning pock-mark, is interesting. The purchaser of the 1854 edition of *Cassell's Latin Dictionary* in 896 pages would seem to have obtained excellent value for money, for as its author remarks in his preface:

> 'It [the dictionary] comprehends every word used by the following authors: Cato, Cicero, Caesar, Celsus, Columella, Catullus, Horace,

Juvenal, Livy, Lucretius, Martial, Nepos, Ovid, Plautus, Phaedrus, the two Plinies, Persius, Propertius, Quintilian, Sallust, Seneca, Silius Italicus, Statius, Suetonius, Terence, Tibullus, Tacitus, Virgil, Vellejus, Varro, Vitruvius, Valerius Maximus, and Valerius Flaccus.'

None of these authors used the word *variola*, or any word for pock-mark. The inference is that smallpox did not exist in classical Rome.

If one presses the point by seeking the Latin for pock-mark in the English-Latin section of Cassell's dictionary, sure enough there is *variola*, but with an asterisk that is said to denote words of 'modern origin'. By modern origin is meant medieval Latin.

On this evidence we predicted that there would be no word for pock-mark in ancient Greek, and the classical scholars we have consulted assured us that this is indeed the case. Nor is there any description of smallpox in the Hippocratic writings. The Hippocratic descriptions were capable of being very clearly expressed, as the following shows for the case of mumps:

> '. . . Many people suffered from swellings near the ears, in some cases on one side only, in others both sides were involved. Usually there was no fever and the patient was not confined to bed. In a few cases there was a slight fever. In all cases the swellings subsided without harm and none suppurated as do swellings caused by other disorders. The swellings were soft, large and spread widely; they were unaccompanied by inflammation or pain and they disappeared leaving no trace. Boys, young men, and male adults in the prime of life were chiefly affected . . .'

It defies commonsense to suppose that ancient Greek doctors would trouble themselves to observe mumps so carefully and yet would by-pass the deadly disease of smallpox, for which there is no cure or treatment, and which in times past killed some forty per cent of its victims. Survivors are often so seriously disfigured that it also defies commonsense to suppose that classical authors would have made no mention of the disease, if it had then existed.

Lesions in the skin of mummies dating from 1550 to 1100 BC, spanning the time of Moses, have been interpreted as smallpox, although the existence of the disease is not confirmed in Egyptian papyri. Nor is there mention contemporaneously of it in India, while the situation in China is highly ambiguous at that date. Only some fifteen hundred years later does Ko Hung (ca. 330 AD) give a description of a disease that was probably smallpox.

Plague of Athens: Was it smallpox?

A great plague broke out in Athens in 430 BC. The Peloponnesian War, which marked the decline of Athens, had begun a year before. The history of the war down to 411 BC was written by Thucydides with what has been called minute and scientific accuracy. He describes the plague, giving a detailed account of its symptoms. Some have identified the disease with smallpox, but if so there was no spread like fire in dry grass. Sparta would seem to have escaped its ravages, an advantage for the conduct of the war that some authorities consider to have been decisive.

The precision of Thucydides has provided a challenge of diagnosis to doctors down the ages. After reviewing various diagnoses, one commentator remarked:

> 'I have looked into many professional accounts of this famous plague, and writers, almost without exception, praise Thucydides' accuracy and precision, and yet differ most strongly in the conclusions they draw from the words. Physicians – English, French, German – after examining the symptoms, have decided it was each of the following: typhus, scarlet, putrid, yellow, camp, hospital, jail fever; scarlatina maligna; the Black Death; erysipelas; smallpox; the oriental plague; some wholly extinct form of disease. Each succeeding writer at least throws doubt on his predecessor's diagnosis.'

In our book *Diseases from Space*, we accepted this confusion over diagnosis as showing the Plague of Athens to have been a disease unknown in modern times. Now we are inclined to take the suggested association with smallpox rather more seriously, however. What Thucydides (in *History of the Peloponnesian War*, Trans B. Jowett) wrote was:

> 'The season was universally admitted to have been remarkably free from other sicknesses; and if anybody was already ill of any other disease, it finally turned into this. The other victims who were in perfect health, all in a moment and without any exciting cause, were seized

first with violent heats in the head and with redness and fuming of the eyes. Internally, the throat and the tongue at once became blood-red, and the breath abnormal and fetid. Sneezing and hoarseness followed; in a short time the disorder, accompanied by a violent cough, reached the chest. And whenever it settled in the heart, it upset it; and there were all the vomits of bile to which physicians have ever given names, and they were accompanied by great distress. An ineffectual retching, producing violent convulsions, attacked most of the sufferers; some as soon as the previous symptoms had abated, others, not until long afterwards. The body externally was not so very hot to the touch, not yellowish but flushed and livid and breaking out in blisters and ulcers. But the internal fever was intense; the sufferers could not bear to have on them even the lightest linen garment; they insisted on being naked, and there was nothing which they longed for more eagerly than to throw themselves into cold water; many of those who had no one to look after them actually plunged into the cisterns. They were tormented by unceasing thirst, which was not in the least assuaged whether they drank much or little. They could find no way of resting, and sleeplessness attacked them throughout. While the disease was at its height, the body, instead of wasting away, held out amid these sufferings unexpectedly. Thus, most died on the seventh or ninth day of internal fever, though their strength was not exhausted; or if they survived, then the disease descended into the bowels and there produced violent lesions, at the same time diarrhoea set in which was uniformly fluid, and at a later stage caused exhaustion, and this finally carried them off with a few exceptions. For the disorder which had originally settled in the head passed gradually through the whole body and, if a person got over the worst, would often seize the extremities and leave its mark, attacking the privy parts, fingers and toes; and many escaped with the loss of these, some with the loss of their eyes. Some again had no sooner recovered than they were seized with a total loss of memory and knew neither themselves nor their friends.'

Comparison with a modern medical text shows that a tolerable correspondence exists between most of the positive assertions of Thucydides and the characteristics of the severest form of smallpox, the confluent form. Headache, high temperature, smell, hoarseness, ineffectual retching, and thirst are all found in confluent smallpox. The words 'breaking out in blisters and ulcers' implies pustules in the skin rather than a rash or spots on the skin, and this too is correct. Most diagnostic is the passage: 'While the disease was at its height . . . finally carried them off with few exceptions.' Thus in a modern text:

'In fatal cases by the tenth or eleventh day the pulse gets feebler and more rapid, the delirium is marked, there is sometimes diarrhoea, and with these symptoms the patient dies.'

It is also correct that eyes could be 'lost' through subsequent blindness and that fingers and toes could be lost through gangrene, although these complications are apparently not common in modern times. Post-febrile insanity is sometimes met, which could conceivably explain the last sentence of the quotation: 'Some again had no sooner recovered . . .'

Correct further statements appear in later paragraphs of Thucydides. Thus:

'Equally appalling was the fact that men died like sheep, catching the infection if they attended on one another; and this was the principal cause of mortality. ...'

'For no one was ever attacked a second time, or not with fatal result.'

Omissions rather than positive assertions give the main reason for doubting the association with smallpox. There is no mention of the severe lumbar pains which occur at an early stage of the disease, contradicting the sentence: 'For the disorder which had originally settled in the head....'. More important, there is no mention of the very high density of pustules on the face that is the main outward visual characteristic of confluent smallpox. Nor is there mention of the scarring and disfigurement of survivors. It hardly seems credible that an accurate observer would have by-passed the most terrible aspects of confluent smallpox with the single remark of the 'body breaking out in blisters and ulcers'.

In these circumstances it seemed desirable to return to the original Greek of Thucydides, on the interpretation of which we are indebted to Professor Humphrey Palmer of Cardiff University. We understand that the Greek carries the implication of blisters with an initial appearance similar to those sometimes experienced by oarsmen, and that the word used for 'ulceration' would be consistent with the subsequent development of pustules. This is more informative than the English translation, and would accord better with the eruption of smallpox.

The phrase rendered by Ben Jowett as 'most died on the seventh or ninth day of internal fever' has always caused difficulty for translators. Faced with the improbability that death somehow skipped the eighth day, some have taken the liberty of changing nine to eight, 'some died on the seventh or eighth day', thus presuming that the father of history was unable to count! What Thucydides actually wrote was 'most died in the ninth and seventh days . . .' We understand from Professor Palmer that a controversy existed in the late fifth century BC as to whether weeks should be counted in blocks of seven days or nine days. Thucydides may therefore have meant 'most people died in a week plus a week', and that by using two different lengths for the week he intended his statement to be approximate to within about two days. Deaths from confluent smallpox do in fact mostly occur after roughly two weeks. Less frequently, there are deaths after about five days, and these are from heart failure, which is just what Thucydides says: 'whenever it settled in the heart, it upset it'.

If the Plague of Athens was indeed confluent smallpox, several interesting points follow. We really could then conclude that smallpox was otherwise absent from ancient Greece and Rome, since Thucydides and other classical authors write about the Plague of 430 AD as a unique event. Second, it did not spread into the Peloponnese, and so was not like fire in dry grass. Third, it did not maintain itself. Livy writes of it as lasting for four years only.

The Antonine pestilence which spread in the second century AD through Italy and Western Europe has been ascribed to smallpox on the strength of a description by Galen. If the disease was smallpox then, it would seem to have died out in spite of its infectious quality, for there is no mention of it by other writers of the early Christian era (Celsus, Aretaeus).

Smallpox appeared unequivocally in Arabia during the sixth and seventh centuries. The Caleph Yezid who died in 683 AD was said to be pitted because of it. Epidemics described by Gregory of Tours in the sixth century and by the Venerable Bede in the seventh may also have been smallpox. The disease is caused by a virus that has no known host except man, which circumstance, taken with the nature of the disease itself and with the historic evidence, sets severe problems. To maintain itself, the virus (according to the usual point of view) had to be always passing to new victims as the old victims either died or recovered, since a person who has recovered does not continue to harbour the virus. Thus the virus would become extinct should the chain of case-to-case transmission ever be interrupted. The virus would also become extinct in a situation where on the average each victim infected fewer than one other person, for then the number of contemporaneous victims would decrease steadily to zero along the transmission chain. In the opposite situation, where on the average each victim infected more than one other person, the number of contemporaneous victims would grow along the transmission chain. There would be an epidemic, increasing until the virus at last exhausted the supply of susceptible people.

The condition for smallpox to smoulder over long periods of time is not easy to achieve. It requires a precisely controlled state in which each victim infects just one other person, an unstable condition for a highly infectious disease, except perhaps in the following special situation. Imagine the disease to have run through nearly all the population of some city. The survivors have become immune to the virus and they do not harbour it, so that only the remaining low density of susceptibles is involved. If it were not for births constantly tending to increase the density of susceptibles, the disease would die away. But new births maintain the supply of victims so that the disease propagates itself at a steady level. The disease is then endemic in the city, which becomes a 'focus' from which it can spread to other uninfected populations.

A city endemic to smallpox would be exceedingly noticeable, however, since a large fraction of its population would be survivors disfigured by the disease. There could have been no such city known to Greek or Roman writers, otherwise its exceedingly repugnant nature would surely have been remarked upon. Nor does it seem as if there could have been such a city in India or China, otherwise writers there could hardly have failed to leave a record of it. What is needed is a hidden city, a mystery city, as unknown as a distant comet in space.

Changing patterns of disease

Until some 6000 years ago there were no cities. Over much of the greater part of our history the density of the human population was so low that the smallpox virus could never have found a steady supply of victims. Nor could most of the diseases which afflict us today. Whence then have they come? There is, within the usual point of view, no plausible answer to this question. The usual apology of an answer is to suppose that virulent human pathogens must have had some other means of survival in prehistoric times than they have today.

It is popularly supposed that viruses, bacteria, and protozoa are capable of evolving almost instantly to fill any environmental niche that may present itself, and if this were true in a fundamental genetic sense the situation for the usual point of view might not be so bad. But in spite of a veritable torrent of experiments in the laboratory, such changes have not been demonstrated. Changes in the laboratory are of a negative rather than a positive character, as for instance Drosophila (fruit fly) was changed by squirting it with X-rays, thus tearing up its genetic material. Changes of an apparent kind could also be generated in a microbial population as a result of selective pressures impressed on it by the environment. Thus the environment could, under appropriate conditions, sieve out even trace components of an initial genetic mix, giving a semblance of rapid innovative change.

The same is also true for viruses. Under natural conditions pathogenic viruses are supposed to undergo all manner of weird and wonderful changes. Yet a genetically well-defined virus injected into a laboratory animal comes out (after multiplication in the animal) as essentially the same virus that went in. Laboratory workers are at their wits' end to understand the bewildering changes of viruses like those of influenza and the common cold, changes that seem to be taking place with the greatest of ease in the outside world. Nothing ever happens in the laboratory itself, yet pandemonium seems to occur in the local cafeteria and in the home. Like a well-known kind of dream, when you look nothing moves, but whenever you look away for the briefest moment everything changes.

The situation is easily explicable if one supposes that viruses, bacteria, and protozoa are continuously incident on the Earth from space, with a range of biochemical properties significantly wider than the habitats available on the Earth can admit. The minority that succeed in fitting themselves to terrestrial habitats fill all the available niches. Should a new habitat arise, through men beginning to dig coal out of the ground for example, then immediately a life-form already present among the infalling types fills the new niche. The process is essentially instantaneous, and it needs no terrestrial evolution at all.

For pathogens, the pattern is one of never-ceasing change, due in part to the varying immunities of their hosts and in part to an ever-changing distribution of the pathogens themselves. Human populations were exposed to smallpox in the sixth and seventh centuries because it was then that the smallpox virus was incident from space. Classical times were free of the disease because the Earth did not happen to encounter the virus over a millennium from about 800 BC to 200 AD.

A few pathogens are rare visitors to our planet, a possible example being the one responsible for the Plague of Athens. Another was the bacterium or virus responsible for the so-called 'English Sweats', which came in five epidemics between 1485 and 1552. The symptoms, besides sweating, were thirst, nausea, and fever. Unlike smallpox, an attack of the disease did not confer immunity against subsequent attacks. Many thousands are said to have experienced repeated bouts, and perhaps a million people are thought to have died from the disease, which, although apparently first described in London, was present also in Western Europe. The 'English Sweats' disappeared as suddenly as they had appeared, as if miraculously prohibited.

Man is the only known host of the measles virus, and as with smallpox a single attack confers a lifelong immunity. Measles is also a disease in which the virus disappears when recovery or death occurs. Consequently much the same problems exist for measles as for smallpox, particularly as the histories of the two diseases are remarkably similar. There is no description of measles either in Hippocrates or in the Indian medical writings of the sixth century

BC. Since the rash associated with measles is very characteristic, this surely implies absence from ancient Greece and Rome. There is uncertain evidence of its appearance in the second century AD, and more certain evidence from the sixth century. As with smallpox, measles was clearly described by Abu el Rhazes (860-932 AD). Rhazes seems to have considered the two diseases of comparable severity, perhaps indicating that measles was then attacking adults who had not acquired immunity in childhood.

The World Health Organization (WHO) has declared the world to be free from smallpox, as a result of quarantine measures and of an intensive vaccination programme. One can wonder why the world has not similarly been made free from measles. The incubation period for measles is even longer than for smallpox, about 14 days, so that quarantine measures for measles should be similarly effective. An effective vaccine for measles also exists. Faced with the contrast between the eradication of smallpox and the ever-present quality of measles, we would argue that the measles virus is continually incident from space, whereas the smallpox virus was a survivor from incidence in the past. According to the conventional point of view no explanation can be offered, except perhaps to say that the world is too indifferent towards measles for a serious attempt at its eradication to be made. We doubt that this explanation is correct.

The Black Death

Turning now from virus-induced diseases to those caused by bacteria, the most famous example is that of the Black Death, or bubonic plague. Unlike the cases considered above bubonic plague is not primarily a disease of man. The bacterium *Pasteurella pestis* attacks many species of rodent, its first target. Because the black rat happened to live in close proximity to humans, nesting in the walls of houses, the physical separation of people from the black rat was small and could be bridged by fleas, which carried the bacterium from the blood of the rats to the blood of humans. This transfer process was not welcomed by the fleas, who preferred to stay with the rats, quitting them only as their hosts died from the disease.

Although human-flea-human transfer of the bacterium

presumably occurred, it does not seem to have been sufficient to maintain the disease, which died out as the supply of rats became exhausted. The saving grace was certainly not the prayers that were said everywhere in the churches. By an accident of providence the fleas did not fare well on a diet of human blood. The affliction was irrationally thought to be a punishment from God imposed (like the Flood) as a reprisal for human wickedness. Yet it was precisely in the institutions of the Church where the death rate was highest, particularly among monks in the monasteries, crowding together for their prayers.

As with smallpox, bubonic plague has come in sudden bursts separated by many centuries, and there are the same difficulties of understanding where *Pasteurella pestis* went into hiding during the long intermissions. A somewhat ambiguous reference occurs in the Old Testament, at a date of about 1200 BC, when the Philistines are said to have been attacked by 'emrods [buboes] in their secret parts . . .' as a reprisal of God for an attack on the Hebrews. A clear reference to plague occurs in an Indian medical treatise written in the fifth century BC, in which people are advised to leave houses and other buildings 'when rats fall from the roofs above, jump about and die'.

There may have been an outbreak of plague during the first century AD, with centres of the disease in Syria and North Africa, but between the first and sixth centuries there were no known attacks of the disease. In 540 AD, a pandemic involving the Near East, North Africa and Southern Europe is said to have had a death toll that reached 100 million, with more than 5000 dying each day in Constantinople alone. This was the so-called Plague of Justinian, the Roman Emperor of the time.

Bubonic plague would then seem to have disappeared from our planet for eight centuries, until it reappeared with shattering personal and social consequences in the Black Death of 1347-50. Thereafter, the disease smouldered with minor outbreaks until the mid-seventeenth century when for two centuries it seemed once again to have died out, only for it to reappear in China in 1894. In India, it killed some thirteen million people in the years up to the First World War.

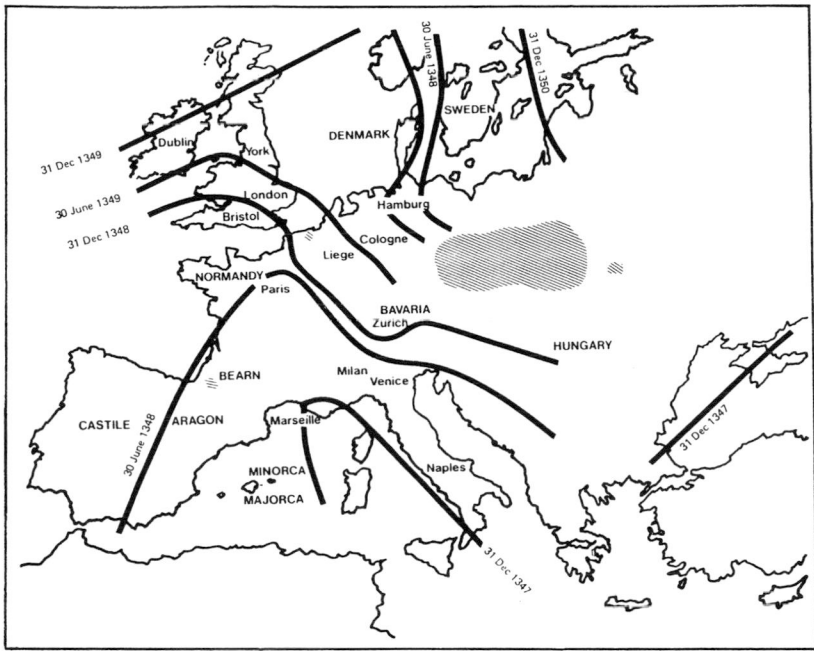

Fig. 6.1. The spread of the Black Death through Europe (after Dr. E. Carpentier).

By working from historical records of the first outbreaks of plague in various population centres, Dr E. Carpentier obtained contours (Fig. 6.1) showing the spread of the Black Death across Europe, except that the one for 31 December 1347 appears to have been drawn to fit a belief rather than from documentary facts. The belief is that the disease was brought into Europe from the Tartars who had besieged the Genoese base of Caffa in the Crimea. The line of December 1347 is said to mark the voyage of Genoese ships back to Italy. One can have some doubt about it, however, since the much better attested contour for December 1348 seems to be headed to intersect the one for December 1347 somewhere in the region of the Danube delta, which from the nature of the contours – the first outbreaks at the various locations – is an impossible condition.

The contours of Fig. 6.1 are interpreted by orthodox opinion as steps in the march of an army of plague infested rats. Humans with the disease collapsed on the spot, and we think afflicted rats must surely have done the same. To argue that stricken rats set out on a journey that took them in six months, not merely from southern to northern France, but even across the Alpine massif, borders on the ridiculous. Nor does the evidence by any means support in detail such an inexorable stepwise advance of the plague. Pedro Carbonell was the archivist to the Court of Aragon, a post that in its very nature could only be held by a person with a clear appreciation of the difference between fact and fiction. Carbonell reports that the Black Death began in Aragon, not at the Mediterranean coast or at the eastern frontier, but in the western inland city of Teruel.

Since it apparently stretches credulity too far to argue that the advancing army of stricken rats also managed to swim the English Channel, it is said that the Black Death reached England by ship. Yet the contours of Fig. 6.1 are of quite the wrong shape for boats to have played a significant role in spreading the disease. If *Pasteurella pestis* had been carried by sea, the earliest contour would be wrapped around the coastline from the Mediterranean to northern Europe, with subsequent contours then filling in gradually towards central Europe. Not only this, but if the bacillus had travelled by sea, the coast of Portugal would have been seriously affected, whereas the evidence is that the plague scarcely penetrated to Castile, Galicia, and Portugal.

There are many descriptions of communities that isolated themselves deliberately from the outside world, many such descriptions from English villages. Yet isolation was to no avail. The Black Death would strike suddenly, and within a week the people in such a community would be just as affected by the disease as everyone else.

What remarkable rats they were! To have crossed the sea and to have reached into remote English villages, and yet to have effectively by-passed the cities of Milan, Liege, and Nuremberg! To have reached into remote villages and yet to have largely spared the shaded areas of Fig. 6.1, especially the extended area in Bohemia and southern Poland. The astonishing reason offered for this behaviour is indicative of the state of mind engendered by orthodox theory. The rats, it is said, disliked the food available to them in

Fig. 6.2. Schematic representation of pathogenic clouds settling at ground level. The patches (shaded areas) cover about one third of the total surface.

these regions.

We remarked above that Indian doctors had noticed already in the fifth century BC the connection of plague with rats. Yet medieval doctors had no such thoughts. It was their overwhelming view that the pestilence had its origin in the air – 'poisoned air' was the widely favoured explanation. It has been fashionable to decry this view as an unsubstantiated superstition, although the state of technical understanding in fourteenth century Europe was higher than it had been at any earlier stage of human history. Indeed, unless one is prejudiced by modern superstitions, the contours of Fig. 6.1 are a clear indication that *Pasteurella pestis* hit Europe from the air. There was no marching army of plague-stricken rats. The rats died in the places where they were infected, just as humans did. By falling from the air, *Pasteurella pestis* had no diffficulty at all in crossing the Alps, or in crossing the English Channel. Remote English villages were hit, however determinedly they sought to seal themselves off from the outside world, because the plague bacillus descended upon them from above; and against an aerial assault all the precautions taken were of no consequence. Milan, Liege, and

Nuremberg went comparatively unscathed because it is in the nature of incidence from the upper atmosphere that there will be odd spots where a pathogen does not fall, as shown in Fig. 6.2. So too did Bohemia and southern Poland escape, even though these areas grow food just as palatable to rats as everywhere else.

6.2 ALLEGED INFECTIVITY OF INFLUENZA AND SIMILAR DISEASES

Let us now turn to modern diseases that are far less serious than the bubonic plague. Influenza, for instance has only a small morbidity attached to it, but it is responsible for a large fraction of absences from work and school.

It is generally thought that acute upper respiratory tract infections are caused by the intake of viral particles that were previously exuded by some other person, or in rare cases by some other animal. Yet little or no evidence capable of standing up to critical analysis has ever been presented in support of this widespread opinion, which appears to have arisen through historical accident rather than through accurate observation and experiment. Following Pasteur's classic experiments on alcoholic fermentation and silkworm diseases it became established that some human diseases arise from the transmission of bacteria from person-to-person, and since in the later decades of the 19th century there was no appreciation of the difference between viral and bacterial diseases the concept of infection by person-to-person transmission became applied to all diseases, a point of view that appeared to gain support when in 1892 R.Pfeiffer (*Dtsch. Med. Wischr.*, 18, 28, 1892) mistakenly implicated the bacillus *H.Influenzae* as the causative agent of influenza. A few epidemiologists, notably Charles Creighton in Britain, continued to protest that the evidence contradicted the rising tide of medical opinion, but in an age when few students had the leisure, affluence and inclination to examine the facts for themselves, the 19th century belief became set rigid in the educational system.

Once a false belief becomes established it is very difficult to get it out, essentially because the system invents supposed facts in order

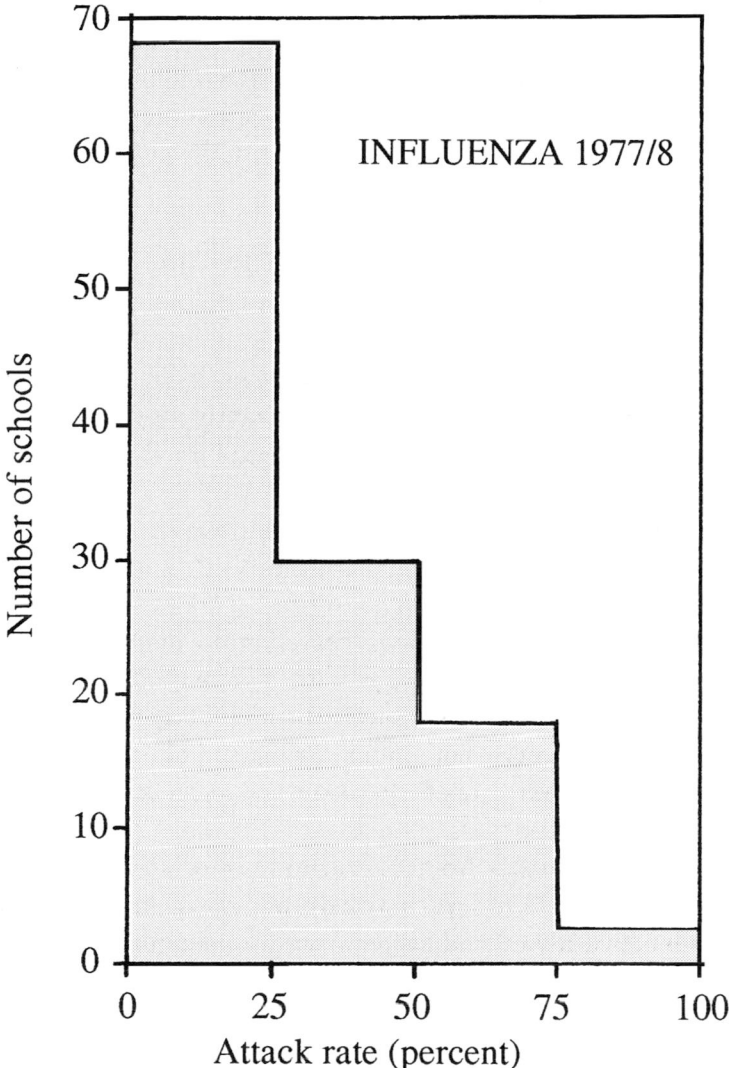

Fig. 6.3. Distribution of attack rates of influenza in the 1977/78 epidemic.

to support it. We had an experience of this process in action following the peculiar epidemic outbreaks of influenza A in the winter of 1977-78, peculiar because of the return of influenza subtype
H1N1 with a variant dating apparently from the year 1950. It is
commonly stated that the person-to-person transmissibility of
influenza is proved by very high attack rates in institutions such as

barracks and boarding schools. Yet a survey of boarding schools in 1977-78 involving a total of more than twenty thousand pupils with a number of victims estimated to be some 8880 for an average attack rate of about 30%, yielded the distribution of attack rates shown in the histogram of Fig. 6.3. In fact, only three schools out of more than a hundred at the extreme upper end of an approximately exponential distribution had the very high attack rates which have been claimed to be the norm.

All the diagnoses involved in these data were made by school medical staffs in advance of our enquiries. Possibly other respiratory infections became associated with influenza in the diagnoses, but since January and February 1978 were months of influenza epidemics, and since children of school age had no established immunity against influenza HlN1, the bulk of the reported cases were very likely of this disease. And even if, in the absence of isolates or serological tests, one were sceptical of explicit diagnoses, the cases were certainly of acute upper respiratory infection, to which just the same remarks and conclusions would apply regardless of the explicit viruses involved. The schools in question were fee-paying, all with boarders sleeping together in dormitories. The degrees of association of pupils in dormitories, classes and at meal times were not much different from one school to another, and if the virus or viruses responsible for the 8880 cases were passed from pupil to pupil, much more uniformity of behaviour would have been expected. Already in Fig. 6.3 we have evidence of great diversity, with a hint that the attack rate experienced by an individual school depended on where it was located, with some schools being in fortunate places and some in unfortunate places.

The alternative to the person-to-person transmission of a virus is that it falls from the air. For semantic convenience we refer to falling from the air as vertical incidence and to person-to-person transmission as horizontal transmission. Although in this section we are concerned to argue the case for vertical incidence as the cause of most acute upper respiratory infections, it is to be emphasised that we are not making this claim for all viral diseases. While we think that all viral diseases arise in the first place by vertical incidence, it is possible for a virus to establish a reservoir in the human population such that the chance p of contracting the associated disease by human contact is greater than the chance q of contracting the disease through vertical incidence. Normalising so that $p + q = 1$, there are the possibilities $p \gg q, p \approx 1; p \cong q \cong 0.5; p \ll q, q \cong 1$. Diseases in the first of these categories are truly infectious and can be moderated greatly through the old-fashioned method of isolating victims. Indeed one could say that it is just those diseases, as for example smallpox, which the medical profession found to be successfully treated by isolation, that constitute the truly infectious category $p \gg q, p \cong 1$. In this chapter we are concerned with the opposite more numerous category, $p \ll q, q \cong 1$, which includes most acute upper respiratory tract illnesses. Data for measles, the discussion of which goes beyond the scope of this section, suggests that measles belongs to the intermediate category $p \cong q \cong 0.5$. We suspect it is the intermediate nature of measles which explains why the medical profession is divided in its opinions on whether the isolation of victims would or would not effectively stamp out the disease. If our assignment of measles is correct, isolation would appreciably reduce the number of cases but would not stamp out the disease. To stamp out a disease q must be strictly zero, requiring that the input of a virus to the Earth's upper atmosphere shall have ceased.

As regards the input of viruses to the Earth's atmosphere, the particles responsible for the strong ultraviolet component of the zodiacal light must have radii of order 30 nm, the scale of viruses. The density of such particles necessary to explain the observed strength of the zodiacal ultraviolet is remarkably high, implying an addition of $\sim 10^4$ tons per year to the Earth's atmosphere, a total $\sim 10^{26}$ particles added annually. This number may be compared with an epidemic of disease in which each of $\sim 10^9$ humans sheds $\sim 10^{11}$ viral particles, for a total shedding of $\sim 10^{20}$ particles. If only a small proportion of the small zodiacal particles are viruses, if only a small proportion maintain viability, and if only a small proportion interact pathogenically with terrestrial plants and animals, the incident number would nevertheless be so vast that it could dominate horizontal transmission, even under extreme epidemic conditions.

It is commonly assumed that viral diseases are caused by the input to a victim of particles that are substantially identical to the output of viruses from the victim. This assumption is normally made for horizontal transmission, but it is not necessary for vertical incidence. All our data and arguments require a causative agent or trigger to fall from the air, but the resulting disease could be caused by the association of the causative agent with dormant viral particles present already in the victim. Or the whole virus could be involved as with horizontal transmission. The evidence we shall present does not distinguish these possibilities. (See Appendix).

To recap: an agent falling from the air may in general be a viable virus or bacterium, but it could also include any form of organic material. Polypeptides and proteins could act as instigators of virus activity or they could be antigenic and induce or suppress immune response. This could be an allergic response as in hay fever, asthma or perhaps various forms of 'auto-immune' diseases such as rheumatoid arthritis. Proteins arriving through vertical incidence could cause direct effects on proteins present in living tissue, as in prion diseases. It is interesting to note the considerable anecdotal evidence that sufferers from arthritis are very susceptible to changes in weather. Hay fever is certainly linked to airborne pollens, but patients frequently note that their condition is not directly related to recorded pollen levels. The causation of asthma is clearly multi-factorial, including house-dust mite and pollution. Antigenic material introduced by vertical transmission could have a significant role, but the hypothesis has not been tested.

Bacterial diseases can also be thought of in terms of the categories $p \gg q$, $p \cong 1$; $p \cong q \cong 0.5$; $p \ll q$, $q \cong 1$, but with the last category less common than for viral diseases, i.e. dominant vertical incidence being less common. One bacterial disease that is difficult to explain except by vertical incidence, however, is whooping cough. Pertussis has for long been known to occur in cycles of about 3.5 years, which used to be explained on the density of susceptibles theory, namely that after children susceptible to the disease become exhausted by a particular epidemic it was then supposed to take about three and a half years for new births to rebuild the density of susceptibles to the level at which a further epidemic would run. Thus the periodicity of this theory should have been a function of

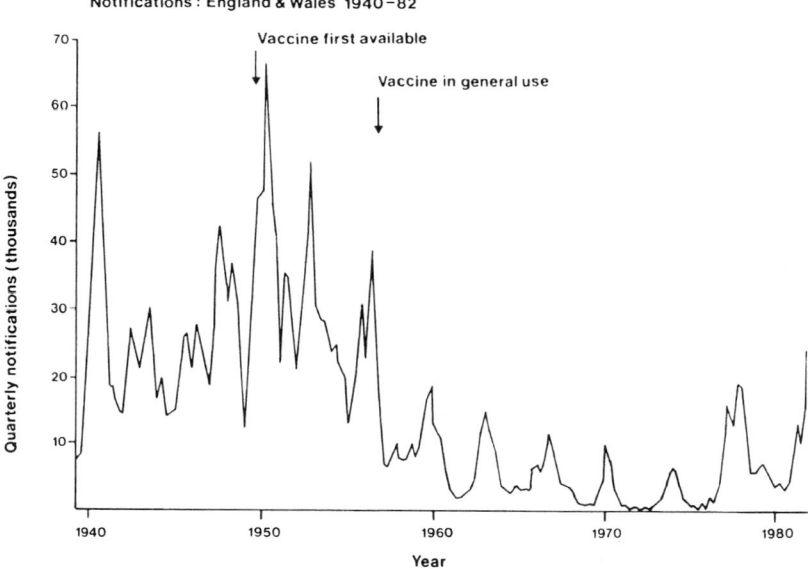

Fig. 6.4. Whooping cough notifications in England and Wales, 1940-82.

population density, with the shorter periods being found in inner city areas of very high density and either long periods or no periodicity at all in lightly populated country areas. But the periodicity was found to be everywhere the same, in town and country alike, and from one country to another. Figure 6.4 shows the record of notifications for England and Wales over the period 1940-82. If the theory had been correct, the sudden reduction in the density of susceptibles brought about in the 1950s by the introduction of an effective vaccine should have greatly disturbed the periodicity, or even destroyed it altogether. Yet the periodicity persisted exactly as before, but with the total number of cases much reduced.

6.3 EVIDENCE AGAINST HORIZONTAL TRANSMISSION

If infectious diseases were propagated from person-to-person, according to the commonly-held view, then people living in high-density city areas should be significantly more subject to disease than people living in lightly-populated areas. From normalised attack rates plotted as a function of population density it would be possible therefore to prove the correctness or otherwise of this point of view. The circumstance that such data do not appear to exist, despite the cogency they would have, is interesting psychologically. Whereas people are avid to collect the slightest scraps of information that support conformist opinion, they are unremitting in their determination not to collect, or even to notice when collected, data which prove the opposite. It really needs no more than the absence of this simple but critical information to see that the commonly held view must be wrong.

One can say in general terms that if any major discontinuity existed between town and country the population at large would easily be aware of it. Moreover, thousands of general practitioners must have had the opportunity to make similar comparisons, without any discontinuity of pattern being emphasised or reported. On a more quantitative level, Figure 6.5 shows data collected by Dr. P. Jenkins, a former Community Health Officer for the City of Cardiff. It gives data covering the three diseases of so-called infective jaundice, whooping cough and measles, obtained quarterly from the heavily-populated Cardiff city area (after normalising to 100,000 population) and from the Vale of Glamorgan, much of which is very rural. Thus each disease in each quarter of a period of three years yields a point in Fig. 6.5. This is except for measles which was so prevalent in one particular quarter that the corresponding point for that quarter could not be plotted without prejudicing the scale of the figure. The one missing point lies on the line defined by the other eleven points, but far away to the right of the figure. Such bias as one can see in Fig. 6.5 goes the wrong way for horizontal transmission. It is the lightly populated Vale of Glamorgan that on a normalised basis appears worse affected. Of course one can always invent the hypothesis that standards of reporting are higher in country practices than in the cities, but we have to doubt quite seriously that this is true. At all events, general experience, together with the data of Fig. 6.5, suggests that there is no marked difference between town and country, as one would expect for vertical incidence but not as one would expect for horizontal transmission.

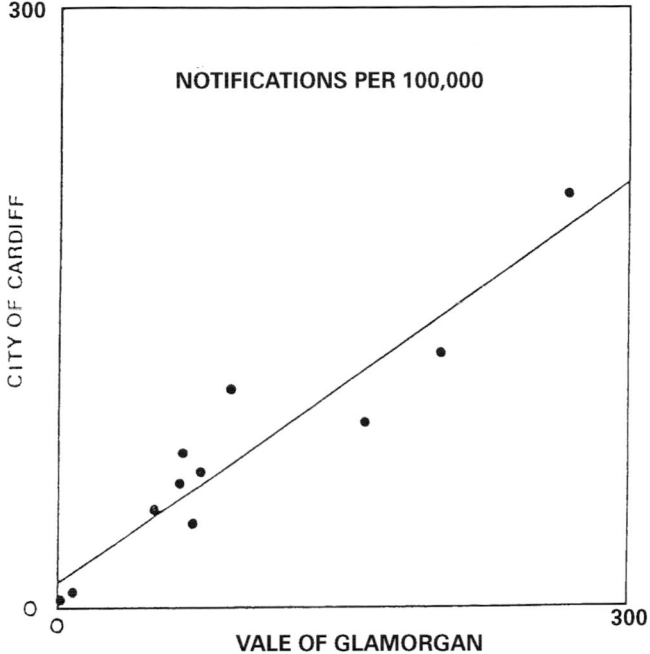

Fig. 6.5 Quarterly incidence of whooping cough, measles and infective jaundice in the City of Cardiff *vs* The Vale of Glamorgan.

Ockham's razor warns us against inventing a 'multiplicity of hypotheses', a warning which some have seen fit to interpret as an edict proscribing the consideration of new ideas. What the warning really means is that we should be on our guard against the invention of a multiplicity of unsubstantiated hypotheses in order to defend conformist views against awkward facts. For example, it is in our opinion an *ad hoc* hypothesis to suppose that city populations possess greater immunity against infectious diseases than rural populations, and to such an extent that the greater exposure which city populations experience with respect to person-to-person transmission is almost precisely compensated by their greater immunity. A similarly *ad hoc* hypothesis would also be required to explain why individuals whose occupations involve exceptional hazards with respect to person-to-person transmission, for instance dentists and cashiers in banks, newsagents and large stores, nevertheless have records of upper respiratory infections that are not noticeably abnormal.

Following in the steps of Charles Creighton, Edgar Hope-Simpson and I. Sutherland (*Lancet*, 1, 721, 1954; *J. Hygiene, Camb*, 83, 11, 1979) were the first in recent years to bring the hypothesis of person-to-person transmission rigorously under the hammer. Hope-Simpson and Sunderland had the idea of defining a set of households by the condition that one member succumbs initially to Influenza A. They then observed the subsequent fates of other members of the households thus defined, finding them to develop no greater proportion of attacks than would be expected for the population at large. Figure 6.6 gives Hope-Simpson's results for epidemics of H3N2 in 1968/69 and 1969/70, shown in the histograms as I and II respectively. Besides the fraction of subsequent cases being normal for the population at large, no well-defined subsidiary peak occurred 3 days after the first cases were reported, as would be expected from incubation if horizontal transmission had been occurring. Hope-Simpson's results have been fully confirmed by P.G. Mann and his collaborators (*J. Hygiene Camb.*, 87, 191, 1981).

A proportion of the fee-paying schools in the survey already mentioned had both day pupils and boarders. The boarders were exposed to close person-to-person contacts for 24 hours a day, whereas the day pupils were only some 8 hours at school, with the remaining 16 hours spent under non-institutional conditions, conditions having fewer person-to-person contacts generally. If there were any substance to the claim of high attack rates in institutions, used to bolster the person-to-person transmission hypothesis, the overall attack rates on boarders should have been significantly higher than it was on the day-pupils. With each of the schools in question represented by a point in Fig. 6.7, the results gave essentially a scatter diagram. Whatever slight bias there is about the 45° line in this diagram disappears for a line of slope 40°, and this is within the expected statistical fluctuation. There are many instances in which the day-pupils experienced considerably higher attack-rates than boarders, a situation that defies the imagination to explain according to person-to-person transmission, for we would have to suppose that after leaving school the day-pupils encountered more seriously infective contacts than were present at

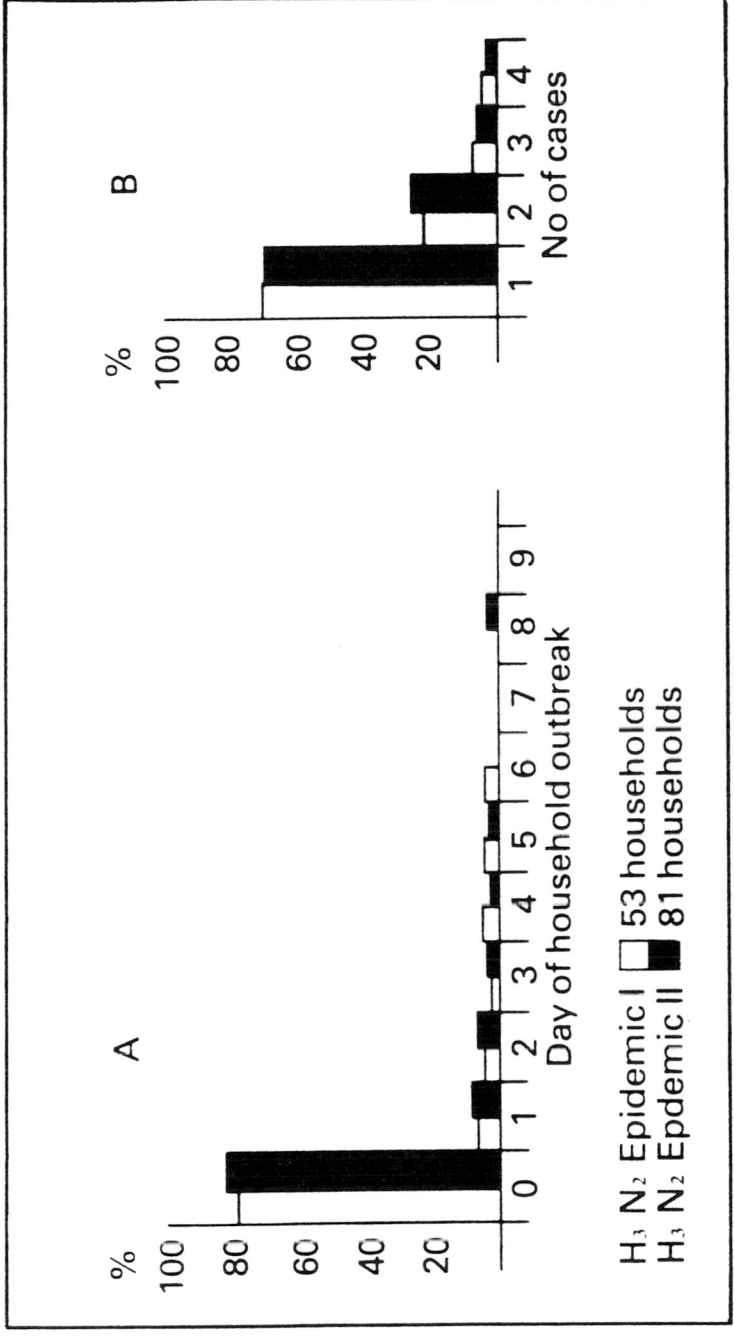

Fig. 6.6. Percentage attack rates in households where one member succumbs to influenza in the epidemics of 1968/69 and 1969/70 at Cirencester, England according to data from Hope-Simpson.

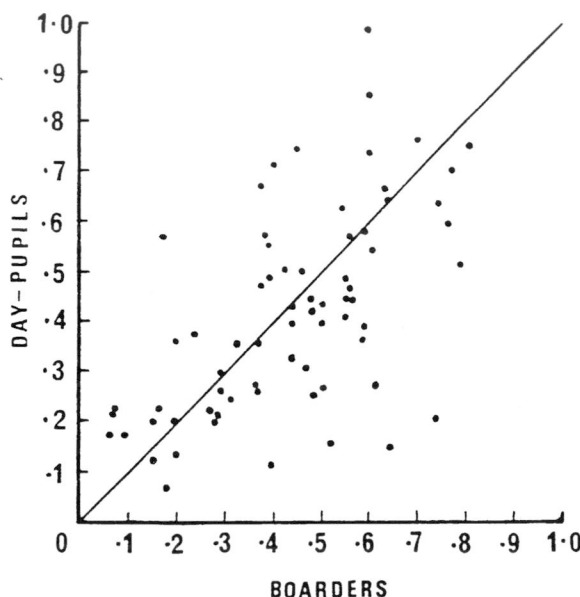

Fig. 6.7. Correlation of attack rates for Day pupils and Boarders for schools which had a mixture of both during the 1977/78 influenza pandemic in England and Wales.

school, and that they did so *systematically* in order to explain high attack rates above 70 per cent for the day-pupils in some of the cases.

In the next section, we shall see that vertical incidence is expected to lead to intricate patchy details in attack-rates, with some localities relatively safe from attack and others relatively dangerous. On this view, schools that happened to be in relatively safe areas would have their boarders staying comparatively safe the whole time, whereas the day-pupils would go out from the school into comparatively dangerous areas and so would experience significantly higher attack rates. And of course the opposite situation would occur for other schools, thereby producing the scatter shown in Fig. 6.7.

Attack rates of around 30 per cent were found most useful for studying variations within school boundaries, since very high attack rates evidently preclude variations being found, while low attack rates gave inadequate statistical weight. Eton College had

441 victims among 1248 pupils, for an attack rate of 35 per cent, with high statistical weight because of the large number of pupils. We were fortunate that Dr. J. Briscoe, the Medical Officer at Eton, had for long been puzzled to understand how his observations could be explained in terms of pupil-to-pupil transmission. Consequently, Dr. Briscoe had collected comprehensive information giving the distribution of victims in some 25 school houses. The houses averaged about 50 pupils each, with about 17 cases expected as the mean number of victims. Such numbers were very suitable for computing standards deviations, with the results shown in Fig. 6.8. Two houses had excess morbidities of 4 standard deviations, two had deficits of about 4 standard deviations, while one house (COLL) had a remarkable deficit of 6 standard deviations. Since pupils in the different houses were mixed in classes and at games these enormous fluctuations from a random distribution are quite inexplicable it seems to us in terms of horizontal transmission. The Eton College results imply that the school was hit vertically by the influenza virus during the night hours, or possibly at a weekend, and that the vertical incidence was patchy enough to distinguish between the locations of the various houses, some houses happening to lie in safe areas and others in dangerous areas. Dr. Briscoe informed us that similar effects had occurred in other influenza epidemics, with the identities of the lucky and unlucky houses being different from the situation in 1978. A patchy vertical incidence would of course not be reproducable in its details from one epidemic to another, so this too would accord with the vertical incidence hypothesis.

We end this section with a somewhat different issue. Younger children have sometimes been found to be more susceptible to influenza than older children. Usually it is not possible to distinguish how far the greater resistance of older children is inherent and how far due to already established immunities. Since no children of school age in 1978 had any established immunity to the HINI subtype, and since some schools in our survey had both junior and senior pupils, it was possible to compare attack rates that gave information largely free of the immunity factor. Results are shown in Fig. 6.9 where each point refers to a school having both junior and senior pupils. These data suggest that inherent resistance has

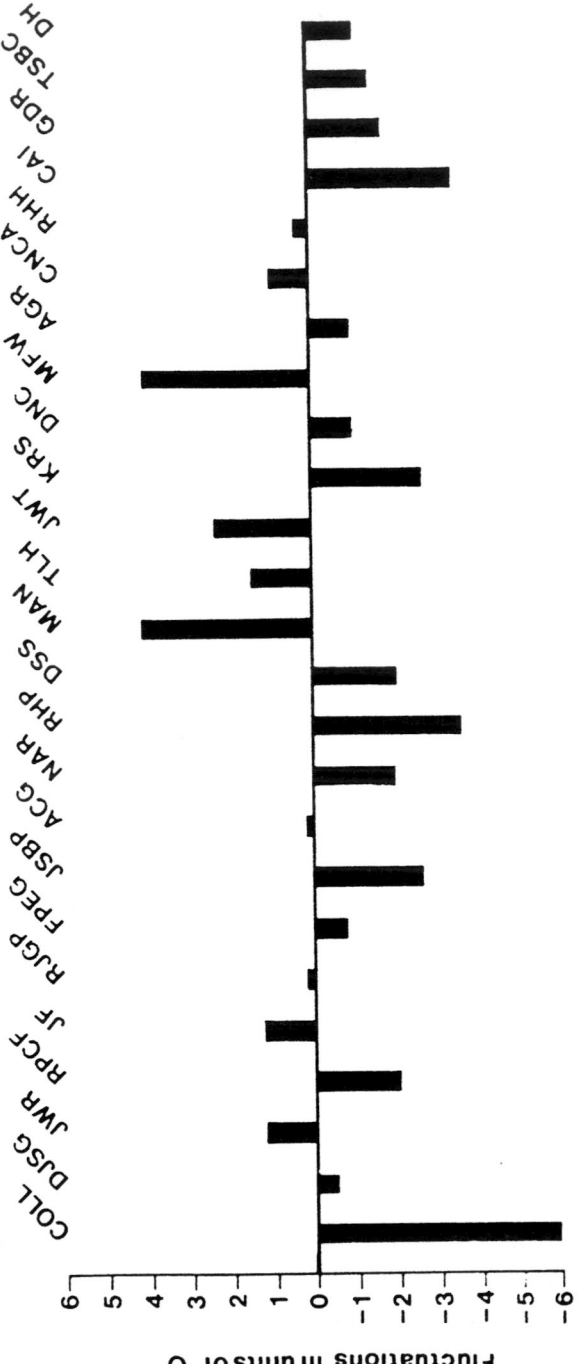

Fig. 6.8. Deviations of attack rates of influenza above the mean attack rate for the 25 school houses at Eton College during the 1978/79 pandemic. The deviations are relative to the standard deviation computed house by house.

ATTACK-RATES

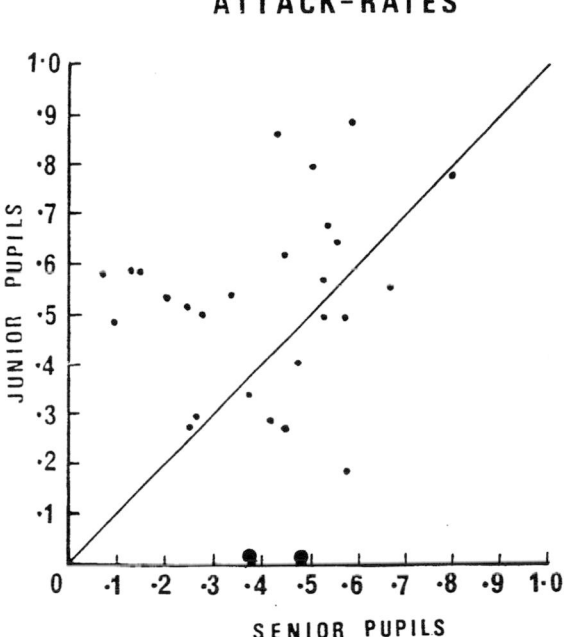

Fig. 6.9. Correlation of influenza attack rates for Junior and Senior pupils (Junior, 5-13 years; Senior ≥ 14) years during the 1978/79 pandemic.

little to do with age, implying that differences observed in other years were related to the immunity factor. Such asymmetry as one sees in Fig. 6.9 about the 45°line (after noting the two very low points marked heavily to catch the eye) would be removed by increasing the 45° slope to a slope of 50°. A slight bias in this sense could have arisen from a minority of cases where the virus subtype was still H3N2 rather than H1N1, with older children having better immunity to H3N2.

6.4 THE VERTICAL INCIDENCE THEORY

According to medieval lore, diseases come from comets, and according to our view this is true, but only in a broad sense. We cannot maintain the dramatic position that ferocious new diseases come from spectacular comets, because for every spectacular comet there are almost certainly very many smaller ones. The smaller comets may not only evaporate more material collectively than large ones but the effect of the Earth crossing almost precisely the track of a small comet would lead to a greater addition of evaporated particles to the terrestrial atmosphere than would a more distant relationship to a large comet, as for instance a distant relationship to Halley's comet.

Support for this position comes from an analysis by Z. Sekanina (*Icarus*, 13, 475, 1970) of some 20,000 orbits of meteors, meteors being small particles typically with sizes ~0.1 mm also evaporated from comets. A minority of the orbits could be associated with known comets but the majority could not. It is perhaps possible to understand both the origin and the demise of the medieval lore in these terms. Over a time-scale of historic length there must have been special situations with the Earth in close proximity to a large comet. If following such special situations serious attacks of disease occurred, particularly if entirely new diseases appeared, their association with the comets in question would be an obvious and natural deduction. But then as other less close comets appeared in the sky, and were not followed by spectacular outbreaks of disease, the belief would be thrown first into doubt and then into ridicule.

Figure 2.5 of Chapter 2 gives a rather mild idea of the complexity of the situation. It shows only the orbits of the so-called Jupiter family of comets. One has to imagine thousands upon thousands of orbits of smaller comets added to this Figure in order to appreciate the real position. At all events, we can see that the Earth is perpetually embedded in a halo of evaporated cometary material, some of the material newly evaporated, without much in the way of exposure to solar ultraviolet light. Because the Earth and the finely-divided cometary material in the halo around the Earth are not comoving, the high terrestrial atmosphere is essential if microorganisms are to make a soft landing here. Speeds relative to the Earth are so high that microorganisms would be destroyed by hitting a hard surface, for instance the surface of the Moon. The maximum size for the safe entry of biological material is ~0.1 mm. This is under the most favourable conditions of speed and geometry. For smaller sizes, those of bacteria and viruses, the situation is not restrictive, however. Particles of the sizes of bacteria and viruses, indeed particles not larger than 0.01 mm, land 'soft' in the high atmosphere, permitting them to retain viability, and subsequently to fall gently downwards as active agents.

The first direct evidence for viral-sized particles in the coma of a comet became available for comet C/1996B (Hyakutake). This comet was found to be exceedingly bright in X-rays during the period March 26-28, 1996, and we interpreted this data[1] to show that a few million tonnes of viral-sized particles were present in the coma and was responsible for scattering X-rays from the sun. A fraction of the virus-like particles would have interacted with the Earth, and almost certainly reached ground level along with winter downdrafts in the northern hemisphere later that year. It is possible that this could have caused some of the increase of viral diseases that were witnessed during December 1996.

6.5 DESCENT THROUGH THE ATMOSPHERE

The lower atmosphere, the troposphere, has a height which varies from about 18 km in the tropics to 10 km in temperate latitudes to 7 km in polar regions. Small particles over the whole size range from viruses with diameters of about 100 nm up to colonies of bacteria descend comparatively quickly from the top of the troposphere to ground-level. The troposphere is a region of falling temperature with increasing height, a physical condition permitting vertical air movements to occur readily, thereby causing water vapour to be carried upwards from the surface regions to essentially the top of the troposphere. The falling temperature with increasing height makes the water vapour supersaturated, but the temperature is not usually so low that the supersaturated water vapour condenses spontaneously into ice crystals. Initially existing nuclei are required for the water to condense around, and the small particles in question provide such nuclei. Thus small particles on reaching the troposphere from above become condensation centres around which

[1]N.C. Wickramasinghe and F. Hoyle, *Astrophysics and Space Science*, **239**, 121-123, 1996.

much larger ice crystals form. Because they are less impeded by air resistance than the original small particles themselves, such ice crystals fall with much increased speeds. As they descend into warmer air the ice crystals usually become melted and the resulting water droplet may either fall to the ground as rain, or it may become partially evaporated and as a smaller droplet remain suspended in the air. Normal precipitation rates are such that this process, often involving repeated cycles of condensation and evaporation of the ice crystals and water drops, serves to wash out the troposphere of small particles in a time-scale of a few weeks. Exceptional conditions may extend the time-scale, however, for example in desert regions or in instances when unusually cold weather is experienced in northern latitudes during January and February.

Above the top of the troposphere, the tropopause, the temperature undergoes inversion through the stratosphere, rising from ∼-55°C at the tropopause to a maximum of ∼3°C at a height of ∼50 km. The physical reason for this temperature inversion is that the ozone in the height range in question, the stratosphere, absorbs solar ultraviolet light very strongly shortward of 3000 A, thereby giving an energy input into the stratosphere. The dynamical effect of the temperature inversion is to inhibit greatly the generally free movement of air such as occurs in the troposphere. Travellers by air will be familiar with the difference between the clarity of the lower stratosphere into which aircraft normally climb and the cloudy turbulent troposphere below. Free air movement in the stratosphere is limited to west-to-east movements along parallels of latitude, of which the most violent are the jet streams. The effect of the free west-to-east movement is to produce a general uniformity with respect to longitude in the stratospheric distribution of small atmospheric particles. If the Earth were smooth at its surface, we would therefore expect any incoming pathogens from space to arrive at ground level at more or less the same times along a given parallel of latitude (although local weather patterns could still introduce modest fluctuations). But because the troposphere has a marked dependence of height on latitude, we would not expect different latitudes to behave similarly, unless the particles happened to be so large that they were able to fall rapidly without much regard to air resistance.

Since the surface of the Earth is actually not smooth, in particular the Himalayas project about halfway up to the stratosphere, the rule of contemporaneous incidence along parallels of latitude is likely to be inappropriate in regions of high relief. Nevertheless, with the exception of the Rocky Mountains of North America there is a belt around the Earth from about 45°N to 60°N where the land is not much above sea-level and where the rule should be applicable. Prague, the capital city of Czechoslavakia lies a little north of 50° and Cirencester in England has a similar latitude within about 1°. Hope-Simpson has noted the similarity shown in Fig. 6.10 between his influenza records for the Cirencester district and the Czech records (*J. Hygiene Camb.* 86, 35, 1981).

A particle of radius 10 micrometres falls through the lowest ten kilometres of the stratosphere at a speed ~ 1 cm s^{-1} and thus takes only a few days to cover what for smaller particles is the slowest part of their downward journey. All particles fall comparatively rapidly through the upper atmosphere above the stratosphere, and then more and more slowly down through the stratosphere. A particle with the size of a typical bacterium, ~ 1 micrometre, falls through the lowest ten kilometres of the stratosphere at a speed of about 2×10^{-2} cm s^{-1} and thus, falls in a time-scale of $\sim 5 \times 10^{7}$ s, i.e. about 2 years. Because there is more of the stratosphere through which such a particle must fall in high latitudes than in the tropics (recalling that the tropopause is higher in the tropics) the slow part of the journey is more extended the higher the latitude. A bacterium falling in ~ 1 year in the tropics would fall in ~ 2 years in temperate latitudes and in 2 to 3 years towards the poles, a situation that is broadly consistent with historical records of the dates of outbreak of the Black Death in various latitudes, as shown in Fig. 6.1. It is interesting to notice in Fig. 6.1 the perturbation of contours of contemporaneous outbreaks produced by the Alps, the mountains of which rise to about half the height of the tropopause.

If a particle of the size of a typical virus, a particle say with diameter ~ 0.1 micrometer, fell under gravity through still air the time-scale for the slowest part of the journey through the bottom

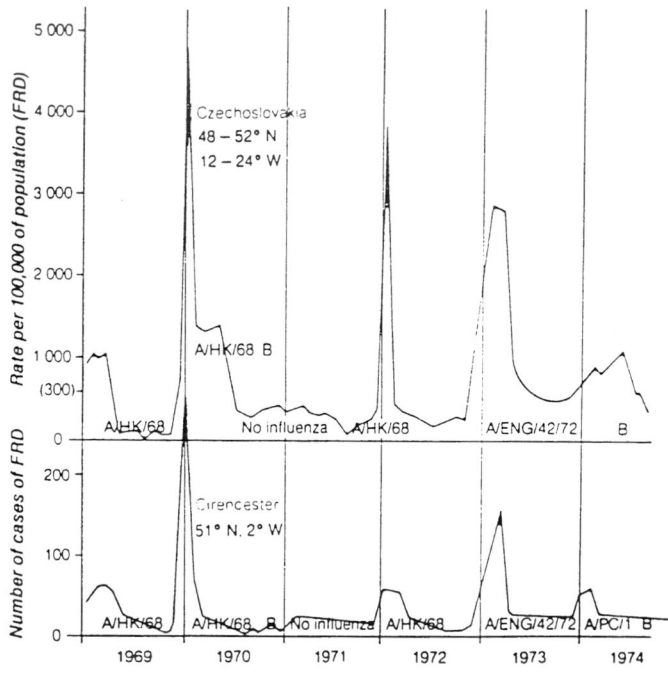

Fig. 6.10. Incidence of influenza in Prague and Cirencester (after E. Hope-Simpson).

ten kilometres of the stratosphere would be $\sim 10^9$ seconds, i.e. about 30 years. This is so slow that other means of descent involving large scale air movements in the stratosphere have to be considered. Although vertical mass movements of air are feeble compared to those in the troposphere, some vertical stratospheric movement takes place despite the inhibiting effect of the inverted temperature gradient. The physical cause of mass stratospheric movements is the equator to pole temperature difference which is available to work a heat engine crossing parallels of latitude, a heat engine that operates more strongly the larger the temperature difference – i.e. much more strongly in winter than in summer. A similar consideration applies also in the troposphere, where an engine crossing parallels of latitude transfers heat from tropical regions towards the poles, again more in winter than in summer. The heat engine in the troposphere is that which we experience in the form of cyclonic storms.

Ozone measurements can be used to trace the mass move-

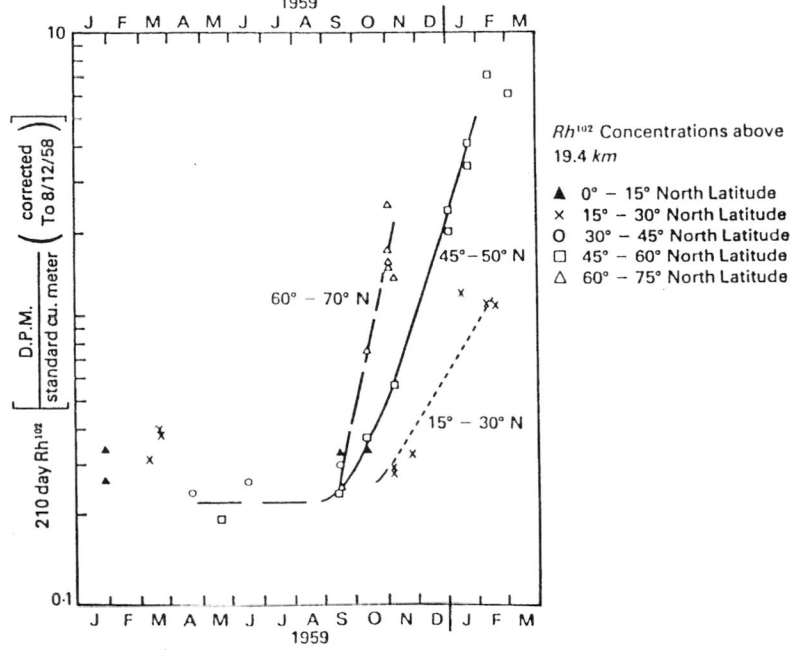

Fig. 6.11. The fall-out at various latitude intervals from the HARDTACK atmospheric nuclear bomb was exploded on 11th August 1958.

ments of air in the stratosphere. Such measurements show a winter downdraft that is strongest over the latitude range from 40° to 60°. Taking advantage of this annual downdraft, individual viral particles incident on the atmosphere from space would therefore reach ground-level generally in temperate latitudes, which therefore emerge from these considerations as the regions of the Earth where upper respiratory infections are likely to be most prevalent, once again on the supposition that the Earth is smooth. The exceptionally high mountains of the Himalayas, rearing up through most of the height range to the stratosphere, introduce a large perturbation on the smooth condition, which may be expected to affect adversely this particular region of the Earth, especially regions lying downwind of the Himalayas, particularly China and S.E. Asia. In effect, the Himalayas are so high that they could act as a drain-plug for most of the viruses incident on the atmosphere at latitude ~30°N, the large population of China being inundated by this drainage effect, making China the quickest and worst affected region of the

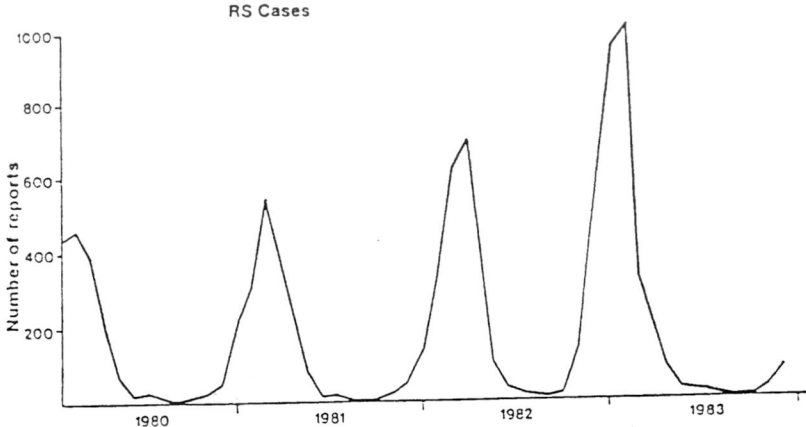

Fig. 6.12. Incidence of RS infections in England and Wales (CDR reports).

Earth. Concomitantly, other parts of the Earth at ~30°N should be largely free of viral particles, unless it happens that such particles are incident as components within larger particles.

A direct demonstration that the general winter downdraft in the stratosphere occurs strongly over the latitude range 40° to 60° was given by M.I. Kalkstein (*Science*, 137, 645, 1962). A radioactive tracer, Rh-102, was introduced into the atmosphere at a height above 100 km and the incidence of the tracer was then measured year-by-year through airplane and balloon flights at altitudes ~20 km. The tracer took about a decade to clear itself through repeated downdrafts of the form shown in Fig. 6.11. Noting that the ordinate scale is logarithmic, the incidence of the Rh-102 is seen to be much greater in temperate latitudes than elsewhere, with the period January to March the dominant months.

The observed incidence of a radioactive tracer agrees closely with the well-known winter season of the viruses responsible for the majority of upper respiratory infections, including influenza. Figure 6.12, (taken from *Communicable Disease Report* CDR 83/49, Public Health Laboratory Service, Colindale, UK) shows the year-by-year incidence of respiratory syncytical virus (RS), demonstrating a remarkable temporal concurrence with the radioactive data of Fig. 6.11. How we wonder is the almost clockwork regularity of RS infections to be explained otherwise? Unfortunately so little has been understood of the mode of attack of so-called infectious diseases that almost any form of hypothesis has come to be accepted in the past as an answer to questions of this sort. The truth is that, although the world may be extremely complex it is nevertheless extremely precise, with explanations every bit as clear-cut as that of the quantum mechanical analysis of the energy levels of the hydrogen atom being ultimately available for every phenomenon we observe.

There has been much recent interest in the appearance of clusters of disease caused by relatively common bacteria that suddenly and mysteriously acquire enhanced virulence. A case in point is meningococcal meningitis. This disease has a sporadic low level of incidence throughout the year in the UK, but significant increases of incidence are noted during the winter months. A mystery associated with this disease is the appearance of geographically localised clusters that could persist for weeks, months or even years. Although the bacterium *Neisseria meningitidis* that is isolated from patients is indigenous to the human population, an agent that transforms this bacterium into a potentially lethal variant could have an incidence from space that could be highly localised.

The conventional wisdom is that the disease is propagated only through intimate personal contact, by means of 'droplet infection'. Yet in all well-documented instances of clusters of the disease this evidence is sadly lacking. In an outbreak that occurred in a Cardiff University hall of residence during November and December 1996, the confirmed cases (including 2 deaths) were apparently not in social contact with each other. Nor were the Cardiff cases linked to other cases reported up and down the UK at almost precisely the same time. There is little evidence here of direct person-to-person spread. The incidence of the disease seems more to be connected with the location where individual victims lived or worked, rather than the people with whom they were in contact.

Variants of the relatively common meningococcal bacterium *Neisseria meningitidis* Group B or C appear in almost every instance of serious disease. We suspect that an airborne (vertically incident) agent, perhaps a plasmid, could be involved in transforming a normally benign *Neisseria meningitidis* into a potentially lethal variant.

If this is so, the primary 'culprit' of meningococcal outbreaks should be seen as the plasmid rather than the bacillus.

A generally similar story might relate to the recent outbreaks of *E.coli* poisoning. The bacterium *E.coli*, like *Neisseria menengitidis*, is a normally harmless micro-organism. An airborne plasmid could be responsible for changing this into one of its lethal variants, *E.coli 0157*, such as had caused havoc in Scotland in November/December 1996, and which was involved some weeks earlier in similar outbreaks in Japan, Canada and North America.

6.6 DETAILS OF VERTICAL INCIDENCE

It was remarked above that the world is extremely complex, and it is interesting to notice how true this can be. Even for the first steps in the acquisition of small cometary particles by the Earth, the Earth could cut just once on a unique occasion through a trail of evaporated particles, or it could cut periodically or irregularly at both closely and widely spaced time intervals. Besides which a pathogenic agent could be carried by a distribution of particles with varying sizes that descended through the terrestrial atmosphere in quite different intervals of time according to the discussion of the previous section.

To recapitulate, particles with sizes $\sim$10 micrometres fall under gravity everywhere over the Earth in only a few weeks. Particles with sizes $\sim$1 micrometre fall in a few years, sooner in low latitudes than in high, while particles with sizes $\sim$0.1 micrometre have a winter season from about December to March (in the northern hemisphere) when they are carried down through the lower stratosphere by mass movements of air, a process that occurs dominantly over the latitude range $\sim$40° to $\sim$60° and which would lead to a cyclically regular disease pattern like that of Fig. 6.12. To these previous considerations we must note that particle sizes can change not only due to water drop formation in the troposphere but to the aggregation of one particle with another produced by sticking effects, should the particles become coated by acid as they pass through a sulphur layer of volcanic origin at heights of $\sim$20km.

Thus time intervals could be appreciably affected by the varying exudations of SO_2 from volcanoes, as well as by the already mentioned effects of high mountain ranges. Finally, we have to consider the complexities that can arise at ground-level itself, complexities giving rise to the local variability in attack rates of a disease such as for Eton College in Fig. 6.8.

The epidemiologist observes the net outcome of all these complexities, with the situation so scrambled together as to present an almost impossible problem of unscrambling, at any rate when the situation is treated empirically. Only with the aid of a model allied to observation can progress be made, as for instance the model of the winter downdraft in the stratosphere (Fig. 6.11) leading to an understanding of the pattern of RS infections (Fig. 6.12). It is a further consequence of this model that similar effects should occur in the southern hemisphere, but six months different in phase because of the alternation of summer and winter in the two hemispheres. We do not have comparable data for RS infections, but influenza behaves similarly and data for influenza confirms the prediction of the model. Thus Hope-Simpson noted the phase variation in the occurrence of influenza across the continent of Africa over the period 1950-51, while we ourselves have assembled in Fig. 6.13 data issued by public health authorities in Sweden, Sri Lanka and Melbourne, Australia over periods ranging from 5 years to more than a decade. The predicted alternation between summer and winter is clearly demonstrated in both Northern and Southern hemispheres, with no significant alternation in the tropics. The global inference from the model is thus confirmed. We have chosen to show the data for Sweden rather than Britain, not because there is any important difference between Sweden and Britain, but to bring out the point that the simple physical cold of winter is not a relevant factor. Sweden has a really cold winter, whereas Australia has a clement winter not much cooler than a Swedish summer. If simple exposure to cold were important, the effect would long ago have been demonstrated under controlled conditions in the laboratory, which it has not been.

Influenza A pandemics, following changes of virus type, do not fit the annual winter cycle in the manner of Fig. 6.13. Influenza

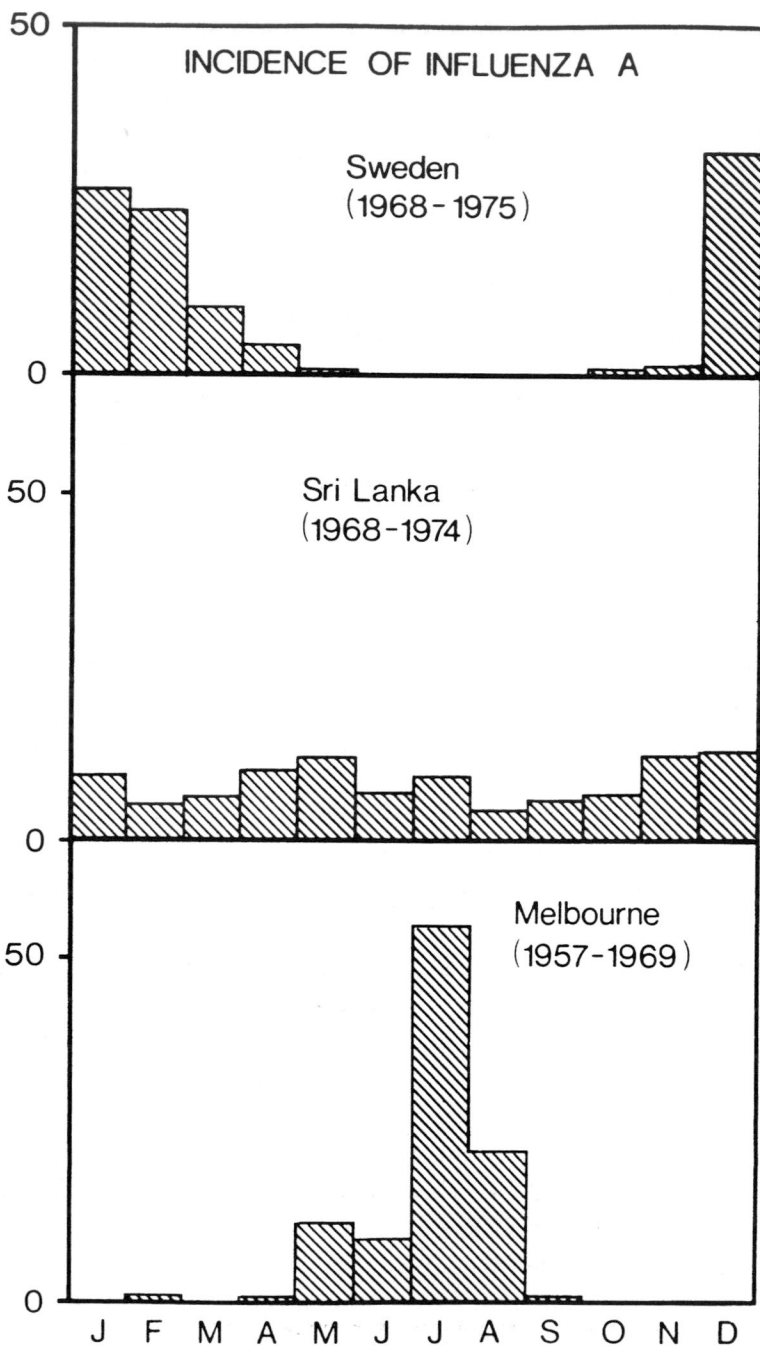

Fig. 6.13. Incidence of Influenza A in three separate countries.

pandemics fit readily into the model, however, with each major influenza type assigned to a separate accretion of virus from space, and with the viral particles being present in larger aggregates as well as individually by themselves. The larger particles fall through the atmosphere under gravity, those with sizes ~10 micrometres in a matter of weeks, those with sizes of ~1 micrometre in a year or two, and those existing by themselves with sizes ~0.1 micrometre taking a decade or more to reach ground-level. It is the latter extended period, characterised by so-called antigenic drift, possibly caused by solar ultraviolet light, that displays the effect of Fig. 6.13, after the earlier accretion of larger particles is over and done with.

Exactly where on the Earth the larger particles, responsible on this view for the initial outbreaks of an influenza pandemic, first reach ground-level is a matter of the vagaries of wind and weather, with the possibility that the larger particles reach ground-level more or less contemporaneously at widely-separated localities. If in accordance with the currently favoured person-to-person transmission hypothesis, one describes the first place of incidence as the 'focus' of the pandemic, the disease would appear to spread from the initial focus to other foci, perhaps separated geographically by large distances, with an amazing and quite inconceivable rapidity. Thus in the 1918/1919 pandemic the first outbreaks occurred within hours of each other in Boston, USA and Bombay, India, an impossibility for person-to-person transmission in days before air travel (see Chapter 2).

Consider next the situation at ground-level itself. It is a matter of experience that we do not normally snuffle raindrops up into the nose or gulp them into the mouth. So if the viruses causing upper respiratory infections fall down through the troposphere inside raindrops or snowflakes it may be wondered how we contract these diseases at all? Rain which impacts the face tends to drip off the end of the nose instead of entering the respiratory tract, possibly the reason for the possession of a nose. But rain does not end because all the water has fallen out of the atmosphere. Rain ends because falling droplets evaporate before reaching the ground. Droplets evaporating immediately in front of one's face, releasing viral particles into the air, would not be harmless because the released particles could then be breathed into the respiratory tract. It is therefore the end of a shower of rain that is dangerous, with the details highly local and irregular, thereby explaining why the incidence of breathable viruses at ground-level is often confined to small patches such as those responsible for the varying behaviours of school houses with respect to influenza, as exhibited in Fig. 6.8. The houses of Eton College which experienced unusually high attack rates were the ones where it happened that pupils were in exposed positions at just the moment when a shower of rain dried up, and conversely for the houses where attack rates were markedly low. It is also apparent why there can be no reproducability from one epidemic to another in the identities of the fortunate and unfortunate houses, the relation of particular places on the ground to the turbulent swirl of falling droplets being essentially a matter of chance, variable over distance scales of the order of the precincts of a school. But not entirely a matter of chance on a scale of miles or on the scale which separates town and country. The efflux of heat from a large city could play a significant role in evaporating water droplets, and if the evaporation occurs high enough above the heads of city people they would be less at risk than country folk, since country folk have no such self-protecting source of heat at their disposal. This may explain why attack rates appear to be somewhat higher in the countryside than in neighbouring cities, as indicated in Fig. 6.5.

Further evidence in the same general direction emerged from a study of the data for influenza in Japan. Japan is remarkable in having high standards of medical documentation and a population density that shows an exceedingly wide contrast from cities to rural districts. Through the good offices of Professor Shin Yabushita we were supplied with annual influenza notification rates in each of 47 prefectures (administrative units) into which the country is divided. We also have estimates of the total population densities in each of these prefectures. Notification rates are only a small fraction of the total case rates, but on the average, a direct proportionality between these two indices would be expected to hold. Table 6.1 shows the relevant data for these prefectures over 5 years. We note that the incidence rates are exceedingly patchy from one prefecture to the next, demonstrating an 'Eton College' type effect occurring over distance scales of some 100's of kilometres. Both high population density prefectures and low population density prefectures show precisely the same effect, an effect which is again markedly in conflict with a person-to-person transmission model.

Table 6.1
Notifications of influenza by prefecture per 100,000

Prefecture number	$N(km^2)$	1970	1977	1979	1980	1981
8 (Ibaraki)	423	6.6	61.4	1.0	16.8	2.5
11 (Saitama)	1580	114.4	1.4	–	0.8	0.1
12 (Chiba)	869	37.4	5.5	–	6.4	4.2
13 (Tokyo)	3600	63.4	2.8	0.4	1.1	0.6
14 (Kanagawa)	2197	51.7	20.5	6.7	5.7	1.4
15 (Niigata)	205	113.3	2.0	5.4	25.2	6.3
16 (Toyama)	225	67.5	312.1	2.5	101.6	4.9
17 (Ishikawa)	217	399.7	95.9	139.1	10.8	0.4
18 (Fukui)	117	375.5	2850.4	–	57.1	264.4
19 (Yamanashi)	189	126.0	19.1	1.0	16.5	0.4
20 (Nagano)	165	5.6	49.7	5.9	3.5	–

Fig. 6.14 shows the influenza notification rate *vs* population density in the prefectures of Japan over a period of 3 years. We note that there is no significant increase of case rate with population density. On the contrary, the reverse seems to be true: the attack rates on the average *decreases* with increasing population density.

It must be rare for snowflakes to evaporate, because at the temperatures $< \sim 0^\circ$ at which snow is able to reach the ground without melting into rain, evaporation rates are low. Hence cold conditions with precipitation falling as snow should on the vertical incidence model be almost free from the danger of upper respiratory infections. Neither is heavy rain a dangerous condition. It is misty, drizzly weather that provides incoming viruses with the opportunity to become dispersed in the air close to ground-level. These expectations agree very well with popular lore, according to which damp is unhealthy but sharp cold weather is healthy. Shakespeare expressed the general lore when he wrote:

... the winds ... have suck'd up from the sea, contagious fogs ...'

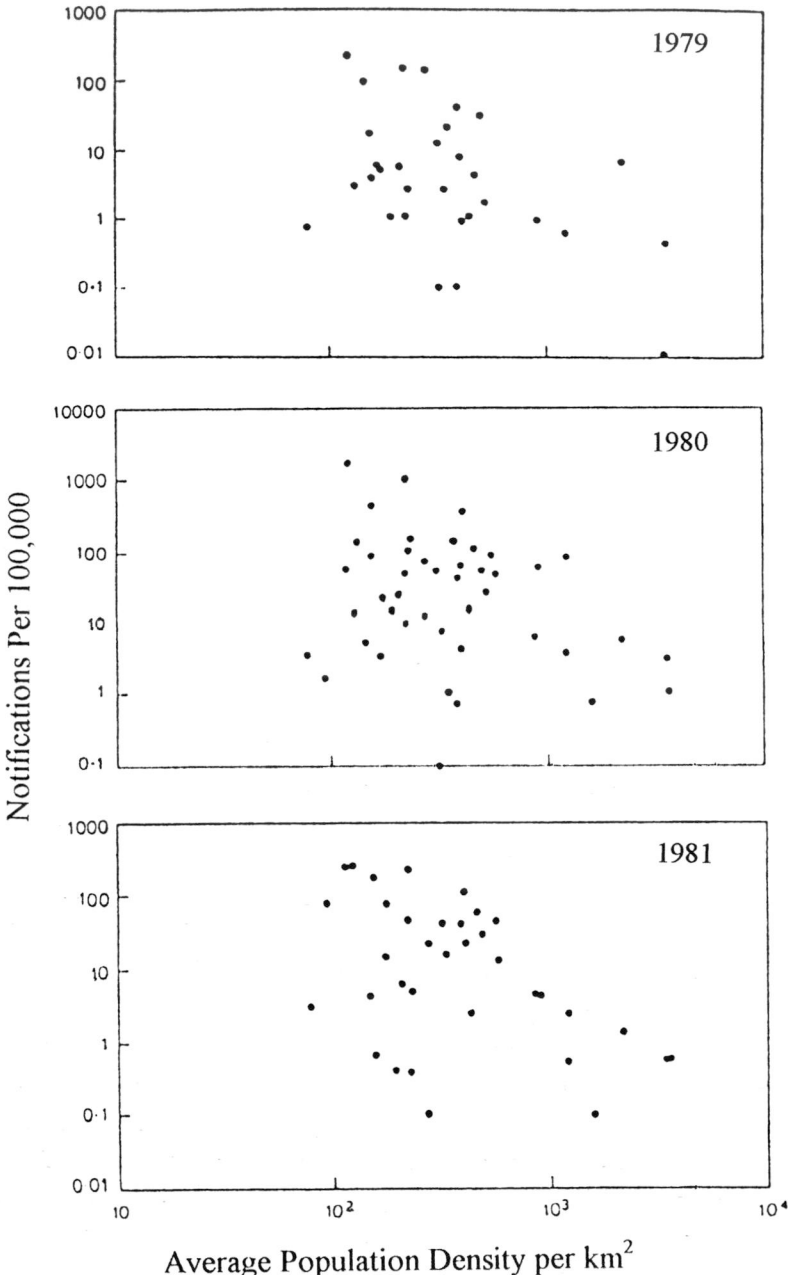

Fig. 6.14. Influenza notification rate vs population density by prefecture in Japan in 1979, 1980, 1981.

In the exceptionally cold weather of January and February 1985 it was widely noticed that Influenza A was essentially absent throughout the world, and that in the U.K. there was an atypical absence of upper respiratory infections generally. We ourselves predicted that once damper, warmer weather set in, as it did in late February, there would be a sharp rise of such infections, as indeed actually happened, with Influenza A at last appearing over the whole latitude belt ranging from the U.S.S.R. to the western states of America. In effect, the exceptionally cold weather held off the infections which normally occur in January and February.

The above discussion has been biased to suit the situation in northern temperate latitudes. If one lives in desert conditions, other factors would be seen to be important. Precipitation in deserts tends to be one thing or the other, either heavy or absent, with little of the damp, drizzly weather of northern latitudes. Viruses falling from the atmosphere would mostly reach ground-level therefore without constituting a serious immediate threat to a desert population. Once on the ground they would largely stay there, however, instead of being washed away in streams and rivers, to be stirred up into the atmosphere again by winds. Windy periods with sandstorms would thus be the times when upper respiratory infections appear, an expectation which according to our somewhat fragmentary knowledge of desert conditions seems to be correct (Prof. M.H.A. Hassan, private communication).

6.7 CONCLUDING REMARKS

On the vertical incidence theory localities of exceptionally high incidence are to be expected at ground-level. Ironically, such regions appear at first sight to give credence to the person-to-person transmission theory, because people living in a locality of high vertical incidence have an impression that they are infecting each other. Likewise whenever any of us contracts an upper respiratory infection after being in contact with an early victim of a similar infection we tend to believe we have 'caught' the infection from the earlier victim. Such perceptions lack quantitative support, however, for whenever they are examined with a little care, discrepancies for the person-to-person theory soon emerge.

Data collected on all scales, ranging from the individual surgeries of general practitioners to worldwide patterns of disease show overwhelmingly that most acute upper respiratory tract infections arise from vertical incidence. The data could be extended greatly at comparatively little effort and cost, as for instance the determination of the attack rates of particular diseases as a function of population density. The reason why such data is not precisely analysed is we think psychological, because it would instantly destroy conformist opinion, which in all branches of science avoids as far as possible confrontations with awkward facts. Yet the benefits to public health from a clear understanding of the cause of acute upper respiratory tract infections could be very great. With the modes of incidence known for various diseases, only quite simple precautionary measures could well serve to reduce morbidities very appreciably, with consequent benefits economically for the community at large as well as for individuals personally.

7

BIOLOGICAL EVOLUTION

7.1 A BRIEF HISTORY OF DARWINIAN THEORY

It is obvious to the eye that remarkable similarities exist between animals and plants which yet do not normally interbreed with each other, between related species as one says, and this fact must have been known for thousands of years. The precise time when the idea first suggested itself to some person that apparently related species really had been related in the sense of being derived from a common ancestral species is not known, although towards the end of the seventeenth century Robert Hooke, who coined the word 'cell' used so widely in modern biology, is said to have been of this opinion. By the latter half of the eighteenth century the evolutionary view had become widespread, particularly in France, to a degree where the systematist Linnaeus accepted it around the year 1770 in order, it seems, to avoid being castigated by his contemporaries as a fuddy-duddy.

The first widely discussed evolutionary theory was published in 1809 under the title *Philosophie Zoologique* by J.B. de M. Lamarck. The theory rested on the postulate that special characteristics acquired by struggles for existence during the lives of parents tend to be transmitted to their offspring. If this postulate had been true, the theory itself would have been logically viable, but many subsequent experiments have shown Lamarck's axiom to be wrong, unfortunately for him.

British naturalists did not begin in the first third of the nineteenth century with a view as wide as the French had held in the eighteenth century, perhaps because of a distrust in Britain, following the Revolution of 1791-94 and the Napoleonic Wars, of everything French. The initial concern of British naturalists was to understand the factors in nature which control the balance of the varieties of a single species. Since the varieties could be observed actually to exist, they were accepted as given entities, requiring no explanation, thus avoiding the pitfall of Lamarck.

It has been said that the first mention of natural selection was made by William Wells at a meeting of the Royal Society of London as early as the second decade of the nineteenth century. The phrase 'natural process of selection' was explicitly coined by Patrick Matthew in *Naval Timber and Arboriculture* published in 1831 (Edinburgh). The idea of natural selection is really no more than a tautology.

If among the varieties of a species there is one better able to survive in the natural environment, that particular variety will be one which best survives. The powers of invention required to perceive this truism could not have been very great.

If evolution leading to the divergence of species from a common ancestor was suspected, and if the concept of natural selection was available, why was the theory of evolution of species by natural selection not under discussion already in the 1830's ? The answer is that it was, as can be seen from the second of two papers pubished in 1835 and 1837 by Edward Blyth (*The Magazine of Natural History*, 1835 Vol.3, pp.40-53; 1838 Vol.1, pp.1-9,77-85,131-141). The first of these papers, *The Varieties of Animals*, is a classic. Besides the clarity with which Blyth addressed his main topic the paper contains passages which foreshadow the later work of Gregor Mendel. In his second paper, Blyth considered the theory of evolution of species by natural selection, telling us in passing that the matter had frequently been dealt with by abler pens than his own. The difficulty for Blyth was that, if 'erratic adaptive changes', as he called the modern concept of mutations, could arise spontaneously in a species, why were species so sharply defined? Why was the common jay so invariant over the large latitude range from S. Italy to Lapland, when surely it would be advantageous for appreciable variations of the jay to have developed in order to cope better with such large fluctuations in its environment ? So quite apart from the unsolved question of the source of the supposed mutations it seemed to Blyth as if the evidence did not support the concept of evolution by natural selection.

The position remained unchanged in this respect for two

further decades until the arrival of a new generation of British naturalists, a position analogous to that which occurred almost exactly a century later in respect of the theory of continental drift. In spite of there being evidence in favour of continental drift, geologists and geophysicists convinced themselves in the 1930's that there were overriding reasons why the theory could not be correct. However, the evidence continued to accumulate to such a degree that by 1960 the situation became inverted. The evidence forced scientific opinion to accept the theory of continental drift, even though nobody understood why continents drifted. So it was with the theory of evolution by natural selection. The evidence forced belief in the theory, even though nobody understood why mutations occur or how the difficulties raised by Edward Blyth might be overcome. So it may also be in the near future in relation to cosmicrobia (Panspermia) as we have indicated in Chapter 2.

The two crucial papers on evolution through natural selection were both written by Alfred Russell Wallace, with titles that left little doubt of their author's intentions. In 1856 he published the paper entitled *On the Law which has Regulated New Species*, and in 1858 a paper entitled *On the Tendency of Varieties to Depart Identifinitely from the Original Type*. Unfortunately for Wallace and for scientific history, he chose to send both papers to Charles Darwin, who had himself been skirting the problem for many years in his personal writings, but who had published nothing nor even communicated his views to his closest friends. With Wallace's second paper available to him, however, Darwin presented a joint communication to a meeting of the Linnean Society and then wrote his book *The Origin of Species* published in 1859. The surprise is that, in spite of the extreme clarity of Wallace's writing, Darwin still contrived to state the theory in a laborious confused way and with an erroneous Lamarckian explanation for the origin of mutations, an explanation which Wallace had himself explicitly eschewed (for a detailed discussion see C. D. Darlington, *Darwin's Place in History* (Oxford, 1959)[1]

If Wallace had published his papers quietly in the *Journal of the*

---

[1]This book appears to exist only as proof copies.

Linnaean Society his views would probably have made as little immediate impact as did the now classic paper of Gregor Mendel. It was the social prestige enjoyed by Darwin, his friends and supporters, that brought the theory of evolution by natural selection forcibly on the world's attention. As always seems to happen when media publicity becomes involved nobody was then interested in precise statements or in historic fact. Writers copied from each other instead of checking original sources, careers were based on the controversy, and attributions became falsified. So did it come about that the theory became known as Darwin's theory, just as two decades earlier the ice-age theory had become known as Agassiz' theory, after Louis Agassiz who propagandised effectively for that theory but did not invent it.

7.2. THE NEO-DARWINIAN THEORY

The work of Gregor Mendel (published in 1866), was rediscovered early in the present century. The work showed that certain heritable characteristics, colours of peas in Mendel's case, were determined by a discrete unit, which was transmitted from generation to generation in accordance with certain simple mathematical rules. Generalising from the small number of characteristics involved in the early experiments, the view soon gained ground that all the gross characteristics of a plant or animal were determined by small discrete units, genes. At the suggestion of W. Johannsen in 1909, the inferred collection of genes for a set of identical individuals in a species became known as their genotype, and the plant or animal to which the genotype gave rise was called the phenotype.

Advances in microscopy pointed to certain discrete objects in the nuclear region of cells, the chromosomes, as the likely site of the genotype. Since the inferred number of genes was much greater than the number of chromosomes, the genes became thought of (correctly, as it eventually turned out) as small structures carried on the chromosomes. Microscopy was not sufficiently refined, however, for individual genes to be distinguished, only the gross forms of the chromosomes. The gross forms for a particular organism became known as its karyotype. Grossly different organisms had readily distinguishable karyotypes, but similar species were often found to have karyotypes that could not be distinguished by the microscopic techniques then available. It was felt, however, that a detailed knowledge of the genes – if it were available – would distinguish between similar species, or even between varieties of the same species. How far this has turned out to be true will be considered in section 7.6.

Experiments of genetic significance in the first half of the century were mainly of two kinds, more complicated examples of the cross-breeding of varieties than those examined by earlier workers, and experiments designed to induce changes in the genotype. Since a gene is a material structure, it was argued, the structure must be changeable by violent means, through irradiation by X-rays for example. It was found possible in some cases to induce changes by such means without destroying viability, although for the great majority of changes viability was weakened in comparison with the original organisms. So genes could be changed, organisms could be altered, mutations could happen it was proved, even though the mutations were deleterious in the overwhelming majority of cases.

Since there could be mutagenic agents in the natural environment, for example the near-ultraviolet component of sunlight and ionizing radiation from cosmic rays, mutations could arise in the wild. Besides which, it is surely impossible to keep on copying any object or structure without an occasional error being made. So quite apart from deliberate mutagenic agents there must be a non-zero copying error rate occurring in the genotype from generation to generation. Here at last, therefore, were the mutations required by the theory of evolution through natural selection. No matter that most of the mutations would be bad, since the bad ones could be removed by natural selection it was argued (erroneously as will be seen in section 7.5). Such then was the position of the neo-Darwinians, who imagined themselves in a stronger position than the biologists of the nineteenth century had been, but the reverse was actually the case. The theory in the form proposed by Wallace would admit of mutational changes coming from anywhere, by additions to the genotype of a species from outside itself, for example through the addition of externally incident genes, as well as by changes to already existing genes. The neo-Darwinians were confined, however, to the already existing genes, and this had turned out to be an insufficient position, as will be demonstrated here and in sections 7.4 and 7.6. The neo-Darwinians boxed themselves into a closed situation, whereas the theory of Wallace could be either closed or open.

The development of modern microbiology from the work of Oswald Avery in the mid-1940's, through that of Erwin Chargaff to the elucidation of the structure of DNA by Francis Crick and James D. Watson, added precision to the concept of the genotype. The genes were sequences of four kinds of base-pair, A-T and its reverse T-A, G-C and its reverse C-G, a typical gene being about a thousand base pairs long. The base-pairs were subsequently shown to be grouped in triplets with each triplet specifying a particular member of a set of 20 amino acids according to the so-called genetic code, the whole gene being a blueprint for the construction of a particular chain of amino acids, a protein or polypeptide. It is through the active chemical properties of its coded polypeptide that a gene expresses itself and is biologically significant (see Chapter 3).

A mutation to a gene could now be seen to consist in one or more base-pairs being changed to another member or members of the set of four possibilities, A-T, T-A, G-C, C-G, this happening to the initial cell at the germination or conception from which an individual of a species was derived. The chance of such a change occuring due to a copying error was measurable, and was found to be generally about 10^{-8} per base-pair per generation[1] – i.e. about 10^{-5} for any base-pair to be changed for a whole gene with a thousand base-pairs. This result was a death knell for neo-Darwinians since it forced evolution according to their views to be a one-step-at-a-time affair, a requirement which both experiment and commonsense showed to be impossible.

Figure 7.1 is a schematic representation of the mode of operation of an enzyme. An enzyme is a polypeptide which coils into an approximately spherical shape but with a highly specific site at its

---

[1]But up to a factor 10 bigger than this in especially sensitive genes.

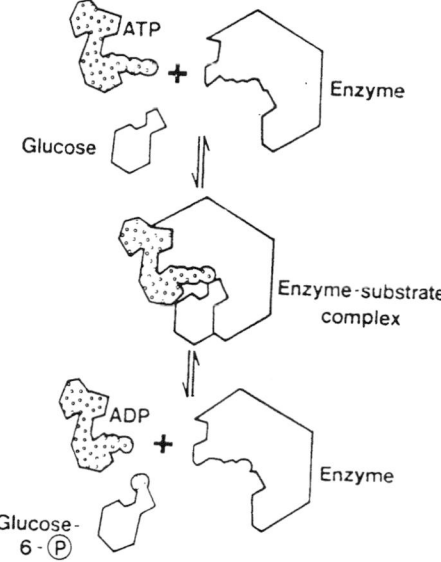

Fig. 7.1. Enzyme action. formation of an enzyme-substrate complex, followed by catalysis.

surface, a site shaped to hold the chemical substances in the reaction which it catalyses, chemical substances existing in many cases outside the biological system itself, chemical substances which do not evolve with the system. This fitting to the shapes of externally defined substances is a constraint an enzyme must meet in order that it should fulfil its biological function. Exactly how many of the hundred (or several hundred) amino acids in the polypeptide chain of an enzyme must be explicitly defined in order that this shape criterion be satisfied is a matter for debate, but the number cannot be trivially small. If it were so, there would surely be far more variability of structure in the enzymes found catalysing the same chemical reaction in bacteria, humans, and in a potato. The number of amino acids in an enzymic polypeptide chain that cannot be changed without destroying the function of an enzyme is probably at least a half and may in some cases be considerably more than a half. This demands that hundreds of base-pairs be appropriately placed in the gene which codes for the enzyme. If one is given an initial situation in which these requisite base-pairs are already correctly placed, well and good, but if the requisite base-pairs are not correctly placed initially, it is essentially impossible that copying errors will ever lead to a functioning enzyme. The difficulty is that all the key base-pairs have to come right simultaneously, not one-at-a-time, because there is nothing to hold individual base-pairs right until the whole lot are right. Every $\sim 10^8$ generations the key base-pairs are randomly shuffled, with the consequence that as some come right others go wrong. The chance of n requisite base-pairs happening to come right at each random shuffling is 4^n so that with $\sim 10^8$ generations required for a shuffling the number of generations needed for a mutational miracle leading to a functioning enzyme to occur is $\sim 10^8$. 4^n, which for n of the order of a hundred is a lot of generations. But not too many for the orthodox neo-Darwinians, who know their theory to be right by some kind of revelation, and who therefore are not embarrassed to offer the most unlikely proposals in its defence.

The standard simple model of genetics may like to tell us that the phenotype is the result of direct programming by the genotype. But consider the information in the structure of the brain alone. There are 10^{10} neurones in the brain with on the average 100 connections each. Thus there are 10^{12} connections to be specified. But there are only about 6×10^9 base pairs in the DNA of the human chromosomes.[1]

There is simply not sufficient information storage for direct linear programming. It seems more likely that the phenotype is the result of the interaction between the genetic programming (including the information in the nuclei, cytoplasm and mitrochondria) and the physical and chemical laws. The programming must include commands for simple iterations. The resulting phenotype occurs due to interaction of genotype, other elements of the fertilised ovum (including mitochondria) and the environment.

It is also important to note that whole viable genetic material can be imported into bacteria from other bacteria, some of which could arrive from space. This occurs, for example, in the development of resistance to antibiotics. Resistant strains can pass genetic material to other bacteria *via* plasmids and the other bacteria can be completely different species.

[1]Jack Cohen and Ian Stewart in *The Collapse of Chaos*, Penguin US, p.181.

7.3. PUNCTUATED EQUILIBRIA OR PUNCTUATED GEOLOGY?

If it were possible to circumvent the criticism of neo-Darwinism given at the end of the last section, arriving at the complex structures of genes several hundreds of base-pairs long by mutations that obtained correct pairs one at a time, with natural selection somehow holding each pair fixed as it came right, evolution would necessarily have to proceed in a very large number of tiny steps, hundreds of steps for each of tens of thousands of distinct genes. There would be two ways to support this point of view. If both worked out well one would be obliged to respect the neo-Darwinian position, but both ways turn out badly, as the argument given at the end of section 7.2 warns that they inevitably will. One way would be to demonstrate the mathematical validity of a small-step genetical theory (to be discussed in section 7.5 below) and the other would be to obtain direct evidence from the palaeontological record showing that markedly separated stages in an evolutionary chain are linked by many intermediate small steps. So far from this being found, new species arise abruptly in the palaeontological record, forcing the neo-Darwinian theory again onto the defensive in exactly the place where it might hope to be strongest if it were true.

Defensively, it has been pointed out (for example by T. H. van Andel, *Nature*. 294, 1981, 397) that present-day sedimentation rates, if maintained throughout geological history, would have resulted in greater depths of sediments than are in fact found from the various geological periods, implying it is argued either much erosion of sediments, in which case the fossil evidence has been largely destroyed, or it might have been that there was a cessation of sedimentation over much of geological history, in which case the fossil record would have been established only sporadically. Evolution in small steps could then be made to appear as a sequence of jumps, simply by the discrete manner in which the evolution happens to be recorded in the presently available fossil record.

All this might be possible as a defensive manoeuvre, but the argument lacks the force of proof. When a curve is drawn through a number of points, the points themselves need occupy only a small

fraction of the total range of the abcissa — what matters for constructing a curve is that there be enough points and that they be
suitably distributed with respect to the form of the curve itself.
Moreover, sediments are available from many geographical areas,
and gaps in one place can be filled by available sediments in another
place, unless erosion or a lack of sedimentation invariably conspired
to be contemporaneous over all areas. For small evolutionary
changes such a complementary association of different areas might
be considered difficult to achieve but if we are looking for big
changes, as from reptiles to mammals for example, a geological
resource of this kind should be possible. One could see the defensive
argument working in particular cases, but it is implausible to
require it to work in every case, as it would need to do to explain
the general abruptness of emergence of new species.

If, on the other hand, evolution really does proceed in sudden
steps which separate extended time intervals of near-constancy,
punctuated equilibria as such an evolutionary process has been
called, one would expect to find examples of abrupt changes within
continuous ranges of sediments. The question of whether sediments
were really laid down continuously or discretely in the manner
discussed above, is a matter for the judgements of professional geologists and palaeontologists. If we have understood their findings
correctly, punctuated equilibria exist (for example, P. G.
Williamson, *Nature*. 293, 1981, 437).

Although neo-Darwinians appear to have convinced themselves that they can explain such findings, we are at a loss to understand their point of view. One might attempt to conceive of many
small mutations being accumulated during a time interval of near-
constancy of a species, of the mutations establishing a potential for
sudden change in a species, like the slow winding of a catapult and
of the catapult eventually being suddenly released. But many small
mutations established without regard for selective control would
mostly be bad, and if there was indeed selective control we should
simply be back again with the previous state of affairs, slow evolution in small steps, not punctuated equilibria.

Large advantageous mutations could explain the findings, but
large advantageous mutations requiring many base-pair changes in

the DNA structure of a gene or genes are exceedingly improbable for the reasons already discussed. Large advantageous mutations requiring only a few base-pair changes might be postulated, but this would be to suppose that genes hover on the edge of marked advantage for species without natural selection having established them in such a critical position. In effect, a *deus ex machina* would be implied. In effect, the theory would have become open in the sense of section 7.2, not closed as it is supposed to be in the neo-Darwinian theory. The position then comes close to our own point of view, to be explained in the next section.

Could abrupt changes to a species be caused by sudden geological changes one might ask? Only to the extent that changes in the physical environment produced selection with respect to the already existing varieties of species. We should then be back with Patrick Matthew in 1831 and Edward Blyth in 1835 (section 7.1). Geological changes could release genetic potential in the sense explained in the following section, but geology cannot create genetic potential.

7.4. EVOLUTION BY GENE-ADDITION

The concept of higher and lower animals, higher and lower plants, is widespread throughout classical biology, and it can be given objective definition in terms of greater or lesser degrees of complexity in the organisation and function of living forms. It is safe to say that if the biologists of the first half of the present century had been asked to guess the relative quantities of genetic material present in various life-forms, general opinion would have favoured a strong positive correlation between quantity and complexity of function, the higher the plant or animal the greater the amount of genetic material. Figure 7.2 shows the results of actual measurements, the one part of the figure for animals, the other for plants, with the various taxa ordered generally with respect to complexity of function (A.H. Sparrow, H. J. Price and A. G. Underbrink, in *Brookhaven Symp. Biol*, 23, 451, 1972). Except that procaryotes do have significantly fewer base-pairs than eucaryotes, and viruses have

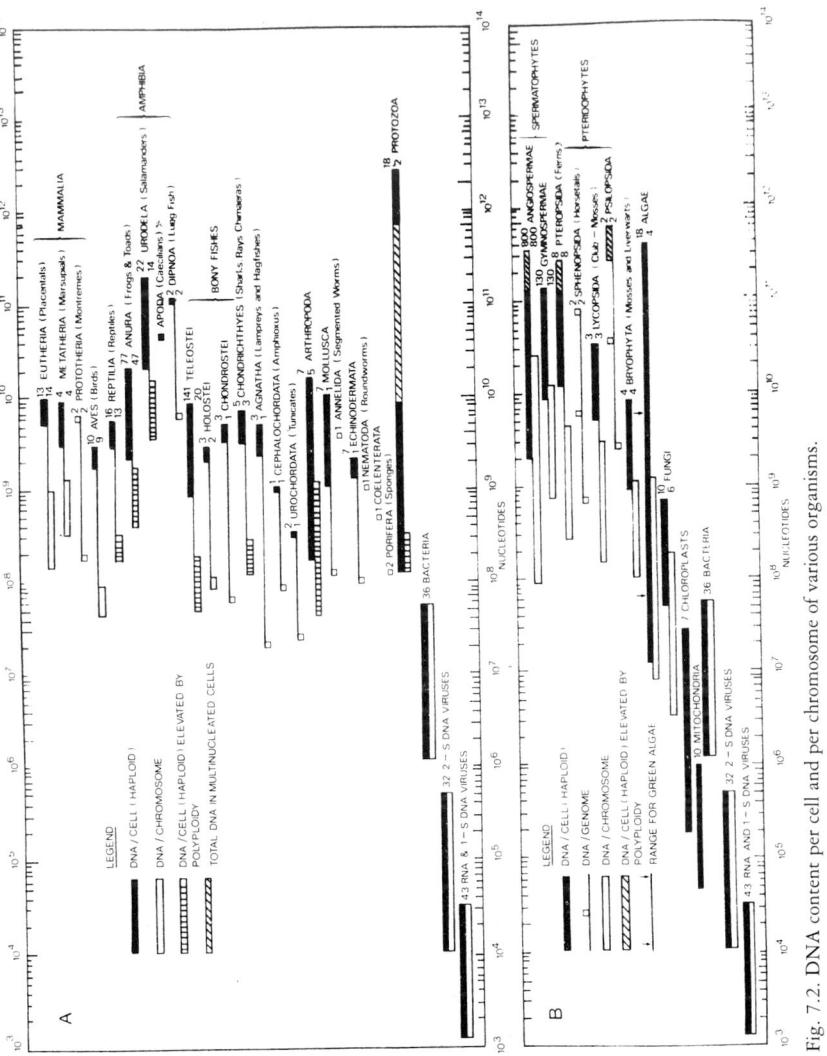

Fig. 7.2. DNA content per cell and per chromosome of various organisms.

still less than procaryotes, the expectation is not borne out. The lungfish easily outclasses the human in the number of its base-pairs. Who would have guessed that the amoeba *chaos chaos* would have had five hundred times more genetic material than the primates?

It might seem odd that the ideas on evolution held by neo-Darwinians have managed to survive Figure 7.2. One might have expected this remarkable new data to have sparked at least one or two revolutionary ideas. The reason for this congealed state of affairs is simply that the usual evolutionary theory explains little or nothing about origins anyway, so that a further mysterious set of facts scarcely makes an already unsatisfactory theory much worse. It is only good theories that can be upset by new facts. A dead horse can take any amount of beating.

Evidence that microorganisms are continuously incident from space was considered in earlier chapters of this book, it being argued that such microorganisms are most readily detected through a component which is pathogenic to terrestrial organisms. Viruses and viroids were considered as well as bacteria, microfungi and protozoa. Some commentators (not professional virologists, at least not to our faces) have claimed that pathogenic viruses cannot be incident from space, for an imagined reason which they believe overrides the many facts which prove otherwise. The argument seems on minimal thought to have the attractive quality of a one-line disproof. Viruses are specific to the cells they attack it is said as if to claim that human viruses are specific to human cells. While a minority of human viruses might be said to be specific to the cells of primates, most human viruses can actually be replicated in tissue cell cultures taken from a wide spectrum of animals, some indeed outside the mammals entirely. The proper statement therefore is that viruses are generally specific to the cells they attack to within about 150 million years of evolutionary history. Actual diseases tend to be specific to particular species it is true, but this is not the same question, which appears to be where confusion has arisen in the minds of some critics. The ability of a virus to produce a clinical attack of disease in a multicellular plant or animal involves the special physical structure and the particular immunity system of the lifeform under attack, and possibly other factors also, all of which are irrelevant to whether the virus can attack individual cells.

If we had knowledge that evolution was an entirely terrestrial affair then of course it would be hard to see how viruses from outside the Earth could interact in an intimate way with terrestrially evolved cells, but we have no such knowledge, and in the absence of knowledge all one can say is that viruses and evolution must go together. If viruses are incident from space then evolution must also be driven from space. How can this happen? Viruses do not always attack the cells they enter. Instead of taking over the genetic apparatus of the cell in order to replicate themselves, a viral particle may add itself placidly to one or other of the chromosomes. If this should happen for the sex cells of a species, mating between similarly infected individuals leads to a new genotype in their off-spring, since the genes derived from the virus are copied together with the other genes whenever there is cell division during the growth of the offspring. Viroids, consisting of naked DNA and perhaps representing only a single gene, penetrate easily into cells, and their augmentation of the genotype may well be still more important than the addition of viruses.

Genes newly obtained in this way may have no evolutionary significance for the plant or animal which acquires them, and for the majority of new genes this would quite likely be so, because each life-form will tend to pick up a random sample of whatever happens to be incident upon it and in the main a gene acquired at random will probably find no useful genetic niche. It will simply replicate with the cells of the life-form in question without yielding a protein of relevance to the environmental adaptation of the species; indeed, if the gene remains unaddressed in the operation of the cell, it will not yield any protein at all. It will remain 'unex-pressed' as one says. So we deduce that many of the genes present in the DNA of every plant and animal will be redundant, a deduc-tion that is overwhelmingly true. Some 95% of the human DNA is redundant. Even higher percentages are redundant in lower animals, which goes some way towards an understanding of how it comes about that a lowly creature may nevertheless have an enor-mous amount of DNA (Figure 7.2).

A gene that happens to be useful to the adaptation of one life-form may be useless to another. Incidence from space knows noth-ing of such a difference, however, the gene being as likely to be added to the one form as the other. So genes that become functional in some species may exist only as nonsense genes in other species. This again is true. Genes that are useful to some species are found as redundant genes in other species. Suppose a new gene or genes to become added to the genotype (genome) of a number of members of some species. Suppose also that one or more of the genes could yield a protein or proteins that would be helpful to the adaptation of the species. The cells of those members of the species possessing the favourable new genes operate, however, in accordance with the previously existing genes, and since the previous mode of operation did not take account of the new genes, a problem remains as to how the new genes are to be switched into operation so as to become helpful to the species. This question is discussed in section 7.6. Here we simply note that, because there is no immediate process for taking advantage of potentially favourable new genes, such genes tend to accumulate unexpressed. As potentially favourable genes pile up more and more, a species acquires a growing potential for large advantageous change, it acquires the potential for a major evolutionary leap, thereby punctuating its otherwise continuing state of little change – its 'equilibrium'. This could very well be the reason why new species appear abruptly, a concept that will be developed further in section 7.6.

7.5. GENETICS IN OPEN AND CLOSED SYSTEMS

According to our point of view, essentially all genetic information is of cosmic origin (see Chapter 3). The information does not have to be found by trial and error here on the Earth, so that mutations in the sense of the base-pair shufflings discussed earlier do not have the positive relevance for us that they have in the neo-Darwinian theory. Indeed, just the reverse. Base-pair shufflings are disadvantageous because they tend to destroy cosmic genetic information rather than to improve it, and this is especially so during the interim period before advantageous new genes are switched into the 'program' of a species, before they become protected from serious deterioration by natural selection. In neo-Darwinism on the other hand, systems are closed, they start with no information and seek somehow to find it, whereas open systems start with high-grade genetic information which it is important for them *not* to lose.

For this latter point of view the base-pair copying error rate should be as low as possible, while for the neo-Darwininians it needs to be high if the requisite sophisticated information is ever to be found, just as the monkeys with their typewriters need to work exceedingly fast if they are to arrive within even a cosmic time-scale at the plays of Shakespeare. The copying error rate is in fact very low, DNA is very stable, clearly supporting the position of the previous section, not that of the neo-Darwinians.

Since many people think neo-Darwinism to be established beyond doubt, and the questioning of it an act of sacrilege, it is worth leading that theory to the knacker's yard yet again, which will be done in the present section. We shall now show that even within its own postulates neo-Darwinism is self-contradictory. At the end of section 7.2 the neo-Darwinian theory was shown to require each important base-pair of every gene (initially not correct) to be held by natural selection when it eventually becomes miscopied to the correct form. The 'discovery' of genes has to be a one-step-at-a-time process. Otherwise there is no possibility worth speaking about of all the many base-pairs coming to their required forms simultaneously. If neo-Darwinism is to be consistent with the detailed structures of genes it is therefore essential that evolution proceeds in very many small steps.

This need to proceed in small steps was already guessed by mathematical geneticists in the first quarter of the present century (e.g. R. A. Fisher, *The Genetical Theory of Natural Selection*, Oxford, 1930). Looking back at this old work it is surprising to find advantageous results for the neo-Darwinian theory being claimed, when even quite easy mathematics shows otherwise, especially as the claimed results were an affront to commonsense. When a mutation is small, its effect on the performance of an individual is so marginal that it scarcely affects the number of offspring born to the individual. Is natural selection really so powerful that in such marginal situations it can stamp out the flood of slightly negative mutations while preserving the trickle of slightly positive ones? Commonsense says no, and commonsense is correct, as we shall shortly demonstrate.

The remedy of R. A. Fisher was to postulate that small negative mutations are not more frequent than small positive ones, but this supposition also defies commonsense, because it is a matter of experience that complex organisations are much more likely to develop faults than they are to find improvements, a view well-supported by modern molecular biology. If the identities of only a hundred base-pairs per gene are important for an animal with 100,000 genes, there are ten million ways at each copying of going wrong. With an error probability of $\sim 10^{-8}$ per copying per base-pair, the chance Q of a significant deleterious mutation occurring per generation per individual is $Q \cong 0.1$. For a breeding group with N members, the number of deleterious mutations injected into each generation is 2QN, which for a typical breeding group, say $N = 10,000$, gives two thousand deleterious mutations per generation, quite a burden to be carried every few years. The number of advantageous mutations must surely be much less than this.

An example will make the situation clearer. Suppose a printer sets up a page of 400 words with a dozen spelling mistakes among them. A single letter somewhere on the page is changed at random, thereby introducing a small 'mutation'. The chance that such a mutation will make the spelling worse, giving thirteen mistakes, is evidently overwhelmingly greater than that the mutation will just happen to correct one of the initial dozen errors. Except that genetically there are only four letters for a base-pair (A-T, T-A, G-C, C-G) instead of the twenty-six letters of the English alphabet the cases are not unfairly compared, especially as the greater number of letters in the literary case is more than offset by the far greater number of genetic 'words', 100,000 genes, any one of which can go wrong.

Since we have analysed the mathematical problem elsewhere (*Why Neo-Darwinism Doesn't Work*, University College Cardiff Press, 1982) it will be sufficient to quote the main results here. In the case of an individual with an advantageous dominant mutation present on either set of chromosomes write $1 + x$ for the ratio of the average number of offspring produced to the average number of offspring for others without the mutation. Then the fraction of such mutations which natural selection spreads through the entire species is about 2x. Thus for $x = 0.001$, a fairly considerable advantage of 0.1 percent, the chance of a mutation spreading through the species is no more than 1 in 500. It therefore needs some five hundred fairly considerable mutations, each of them likely to be a rare event, before just one is retained by the species. Hence for mutations with x small, natural selection adds up very little that is good.

The trouble lies in stochastics, an effect that was inadequately considered by the early mathematical geneticists. For a heterozygote with respect to a gene of small x there is already nearly a 25 per cent chance that the mutation in question will be lost in the first generation, simply from the random way in which the heterozygote allots one or other of its duplicate set of genes to each of its offspring. In the second generation there is again a chance of about $3/16$ that the mutation is lost. Stochastics consists in adding up and allowing for these extinction possibilities, which greatly dominate the effects of natural selection when small mutations first arise.

For the same reason natural selection by no means removes all that is bad, as classical biologists supposed. For deleterious mutations it is the recessive case that matters most. If for simplicity of argument one takes all recessive deleterious mutations to be equally bad an elegant result can be proved. Subject to the disadvantage factor x being sufficiently small, the rate at which deleterious mutations spread through a whole species is equal to the rate Q of the mutations per individual, just the same result as was proved for neutral mutations (M. Kimura and T. Ohta, *Genetics*, 61, 1969. 763).

If natural selection fails for moderate mutations to add up more than a small fraction of what is good, and if natural selection fails to exclude a damaging fraction of the much more frequent disadvantageous mutations, how can species ever become better adapted to their environment? For small-step mutations they cannot, which is why neo-Darwinism fails genetically, why positively-evolving systems must be in receipt of genetic information from outside themselves, as was discussed earlier. The best a closed system can do is to minimise the disadaption to the environment.

Natural selection works excellently for open systems, since with high-grade genetic information coming from outside a system, advantageous changes have large values of x, with 2x of order unity, so that if such a change occurs for only one or two individuals of a species, natural selection operates to fix the change throughout the entire species. Such major advantageous steps have to occur with a sufficient frequency to more than offset the numerous small deleterious mutations which still produce disadaptation at the rate discussed above. In effect, the situation is a race between uphill jumps produced by externally incident genetic information and the downhill slide of the already existing genes, which natural selection can only moderate but not remove entirely. This produces a highly fluid situation, with species either advancing rapidly or sliding backward towards extinction, as is observed to have happened for the higher plants and animals.

When one looks back at the mathematical geneticists of the first half of the present century, it is clear they approached their work in the complete conviction that the neo-Darwinian theory was correct. As the majority of them saw it, their duty was to explain why a theory known to be correct was indeed correct, a mode of argument not unlike a chemist attempting to work backwards through an irreversible reaction, or like an inept student in an examination trying to work backwards from the answer to a problem to its mode of solution. This wrong-headed approach led somewhat naturally to a prostitution of logic which was mercifully concealed from the public in a haze of mathematical symbols. The irony is that the correct answer was easy to find if only the mathematical geneticists had troubled to look for it in the right direction.

7.6. FAVOURABLE MUTATIONS IN OPEN SYSTEMS

Open systems do not have to find genetic information *de novo*, because they are in receipt of genes from outside themselves. However, newly-acquired genes must lie fallow for a while, since the mode of operation of the cells of the species in question cannot know in advance of their arrival. The sequence of events whereby genes are used may usefully be described as the cell program. What needs to be done therefore to promote evolution in an open system is to alter the cell program to take into its operation new genes which it did not use before. The problem to be considered here is the logic of this situation.

A cell program may be thought of as analogous to a computer program. With computers, the program is something different from data and from the closed subroutines which constitute the backing storage. Computers can be operated on many different programs using the same physical hardware and the same backing facilities; examples of the latter are routines for taking logarithms and integrating differential equations. Something of the same kind almost surely exists in biological systems. Genes for the production of enzymes, haemoglobin, the cytochromes, are examples of subroutines that run across all of biology. It is even the case that genes capable of producing some of these standard products, haemoglobin for instance, exist in life-forms which normally make no use of them, just as standard computer languages like FORTRAN or BASIC contain more facilities than are used in any particular individual program.

In days long ago, before sophisticated computer languages were available, when it was necessary to remain closer to the electronic nature of the computer itself, one was perhaps more keenly aware of the distinction between the logical instructions which constitute a program and the numbers or words on which the program operates, even though both were stored in the computer in exactly the same way, as sequences of bits. Although numbers and logical instructions were similar electronically, you could not use numbers for logical purposes or process your logical instructions arithmetically (a few very slick fellows tried and were sometimes successful, but the tricks of this particular trade were too subtle to have survived into current practice!). As well as numbers constituting data and logical instructions making up the program, something else was needed, a starting-point and an end-point, birth and death.

Do biological systems operate in a similar way? Are the logical instructions constituting the cell program stored as genes, but used quite differently from the genes which code for working polypeptides such as the enzymes? Is everything stored as base-pairs in the DNA, just as everything in a computer is stored in sequences of electronic bits? It is tempting to suppose so, but there are indications that it may not be so. The DNA of a chimpanzee is extremely similar to that of a human. Therefore the scope for producing working polypeptides is essentially the same in the chimpanzee as it is in ourselves. Thus the chimpanzee and the human look like two different programs operating on the same physical hardware, on the same backing storage as one might say. If the different programs were on the DNA we might expect to see less close similarity, less homology, between the base-pair sequencing of the two species, unless program storage occupies very little of the DNA, unless the logical ordering which makes us specifically human and a chimpanzee specifically chimp is in each case rather trite and short. Perhaps the logic of being human is rather trivial, but one prefers not to think so.

A less subjective objection is that DNA seems far too stable to be the source of the cell program. If the cell program were so contained, body cells could be replicated a very large number of times without the program being much impaired, permitting animals to have exceedingly long lives, whereas the evidence shows that the program becomes seriously muddled after only a handful of replications. Recognizing this discrepancy some biologists have argued that senescence is itself a deliberate part of the program, deliberate in the sense that natural selection has prevented us from living long by explicitly stopping the coding of essential working polypeptides. This opinion is to be doubted, however, because wild animals commonly die violent deaths before their time is run, so there is no cause in nature for natural selection to prevent lives from being too long. Yet all animals do show senescence, if artificially protected against violent death most of them even more markedly than we do, indicating that senescence is not artificially contrived. The implication is that storage of the cell program must be ephemeral. It is preserved with reasonable fidelity in gametes, but soon runs down and becomes forgotten, leading to grey hair and the like, as soon as the somatic cells are required to replicate more than about sixty times.

If a person tells you that the telephone number of a mutual acquaintance is 752146 and you immediately commit the number to paper you have it in stable storage, like base-pairs on DNA. But if you seek to remember the number aurally in your head, it will be gone at the first distraction, a knock on the door or a pan of milk boiling over on the stove. This seems to be the way of it with our cell program. Once we have lost it, the thing never comes back, although if it really is retained in our gametes somebody may succeed someday in copying it back into our somatic cells, with interesting sociological consequences.

In spite of these difficulties, suppose for a moment that those who think the cell program is written on the DNA are correct. How would the program actually do something ? Not by merely remaining on the DNA, because DNA by itself is inert. The program would need to be translated into polypeptides and it would be the polypeptides that really did something. So why not let the program be polypeptides in the first place? Or if not the whole program, suppose an essential part of it is in polypeptide form, without there being any reference genes coded on the DNA from which the initial polypeptides can be recopied if they become lost. One might conceive for instance that the initial polypeptides comprise a catalogue of what in computer terminology would be referred to as calling sequences, which is to say some means of determining so-called *introns* for finding important genes on the DNA. Senescence looks very much like the progressive garbling of the entries in such a catalogue, so that we end in old-age by not being able to find more than a small fraction of the genes necessary for vigorous life. All this is relevant to the evolutionary problem set out at the beginning, since the less rigidly fixed the cell program the more readily one can conceive of it being changed. The change needed for an evolutionary step must involve some means of addressing new genes added to the DNA, the genes which supply the potential for an evolutionary leap. This means actually doing something, not just adding DNA blueprints for doing something at some stage in the future. Actually doing something means polypeptides, and doing something new means new polypeptides, which implies a working addendum to the old cell program. Where one now asks is such a working addendum to come from? Only it seems from a virus-like particle.

7.7 THE ROLE OF VIRUSES

When a virus invades a cell it mostly happens that the virus multi-
plies itself at the expense of the invaded cell, which it does by stopping the old cell program and inserting its own program, both
necessary but not sufficient properties for what we are seeking. The
several viral particles thus produced then emerge from their host in
search of still more cells to invade, and so on apparently *ad infini-
tum*. This behaviour is usually viewed as a permissible oddity of
biology, permissible because the virus survives, and survival is all
according to the opinions of neo-Darwinians. Yet mere survival
leaves the virus as a disconnected organism without logical relationship to anything else. Once one admits, however, that logical
relationship is at least as valid a concept as survival, indeed that survival is impossible for any organism without logical relationship,
the situation becomes different. The virus becomes a program insertion with the essential capability of forcing cells to take notice.
Many such program insertions are needed to cope with many stages
of evolution for many creatures, both on the Earth and elsewhere.
Hence many viruses are needed, and even if the entry of a particular
virus into cells is restricted to situations in which the cell program
and the viral program match together in a general way, it will not
usually happen that a virus on entering a cell has precisely the
appropriate program insertion to suit the life-form in question
exactly at its current stage of evolution. There will have to be many
trials before precisely the correct program insertion is found. So
what is the virus to do in the majority of cases where the situation
is not quite right ? Give up the ghost and expire? If it did so, what
about the other creatures somewhere in the Universe that may be
in dire need of its particular evolutionary contribution?

Viruses seek cells, not *vice-versa*. Speaking anthropocentrically,
they have the job of driving evolution. They cannot give up the
ghost and expire, otherwise nothing would happen, the situation
would be as dead as mutton. So they augment themselves by
increasing their number and then they press on, forever seeking to
find the cells where they are needed. As soon as one looks for logical design, the situation immediately makes sense. Besides which,

the infective ability of viruses also plays a crucial logical role. For species with a sexual mode of propagation there is a big question mark as to how an evolutionary leap could ever be possible, because the same leap must occur in at least one male and one female, otherwise the male and female gametes will not match properly, and there will be reproductive trouble in the second generation, if not indeed immediately. Since the probability of an evolutionary leap occurring is small, requiring first a building of a potential for the leap and then finding the correct addendum for the cell program, it would be a poor result if the individual for whom all this happened were then to be sterile. Yet if we need the same improbable sequence for the opposite sex also, the small probability is squared, and moreover, the changed male living in London would then have the problem of finding the changed female living in New York, making such an uncorrelated situation quite hopeless. The solution to this last problem is infectivity. The same changes, all being virus induced, can be infective between individuals in close contact at the same geographical location, and in this case the small probability is not squared, and moreover the similarly affected individuals are automatically together and so cannot avoid finding each other. The logic of an evolutionary leap demands infectivity. Infectivity also explains why after an evolutionary leap the previous line does not persist, since with an evolutionary improvement sweeping through a species like a disease, a negative disease as one might say, the previous line is overwhelmed by the superior adaptation to the environment of the drastically changed creatures. Only in this dramatic way can evolution counter the degenerative effect of the small but steadily occurring miscopying of genes, the downward drag that was mentioned earlier. The above discussion also makes it clear why viruses have to be generally specific to the cells they invade.

In conclusion, we accept here that natural selection is not able to fix in a species more than a small fraction of the infrequent advantageous mutations which arise through the shufflings of basepairs on the DNA, and hence that internal processes cannot improve the adaptation of a species sufficiently to be significant. Only by importing genetic information from outside the Earth can adaptation be improved in an important degree, and the observed facts concerning the evolution of life explained. In this connection there is the intriguing possibility that the bulk of new information for the evolution of higher organisms comes in *via* the mitochondria. Mitochondria may be thought of as bacteria that were first introduced into complex cells/organisms billions of years ago. They undergo progressive changes and this fact has been used to argue that all present humans came from one 'Eve'. But what if progressive waves of superior mitochondria were taken into cells and helped or even drove evolution? There would then be muliple 'Eves' all at the same time, and in diverse geographical locations!

EPILOGUE

How frequently do comets collide with the Earth?

Throughout this book we have seen that our planet and all life upon it must be regarded as being derived from the wider cosmos. The importance of comets in this scheme of things is that they lie literally at our doorstep and can provide a steady input of cosmic material. Even today the Earth picks up cometary debris in the form of fine dust, including particles that have recently been found to contain exceedingly complex organic molecules. The average daily input of such material amounts to tens if not indeed hundreds of tonnes, so it stands to commonsense that their effect cannot be entirely negligible. Inevitably also, cometary bodies must collide with the Earth from time to time. How frequently such direct comet collisions take place, and with what consequences, are matters that we shall now discuss.

The earliest cometary collisions occurred some 4.5-3.9 billion years ago and marked the final stages in the accumulation of material for our planet. As soon as this intense initial phase of collisions came to an end we know that life in the form of micro-organisms also appeared. Thereafter comet-Earth collisions died down to a trickle, but did not entirely cease. At the present time a gigantic shell of comets surrounds the entire planetary system at a distance of about a tenth of a light year from the sun. Incursion of comets into the inner solar system are caused by encounters of this shell with individual stars on relatively short time-scales and with galactic molecular clouds on much longer time-scales. The former process leads to a few 'new' comets a year, comets that hardly ever collide with the Earth; the latter process occurring with a time separation of some 100 million years leads to the injection of huge surges of comets, and consequently results in a much increased collision frequency at such times. One could make out a strong case to support the view that the evolution of life on our planet over its entire geological history has been governed by cometary impacts, impacts occurring with periodicities that ranged from a hundred million years to perhaps as low as 1000-1500 years.

There is now little doubt that the extinction of the dinosaurs and of over 75% of all genera of plant, animal and microbial life occurred 65 million years ago (the so-called K/T boundary) as a result of a cometary encounter. This was a process that was first suggested by us as early as 1978. (This work was discussed by us throughout 1977 and published as a paper in *Astrophysics and Space Science,* Vol. 53, p.523, 1978). Later studies by Professor L. Alvarez and his colleagues showed our ideas to be substantially correct. The distribution of the terrestrially rare element iridium and also of extraterrestrial amino acids in the Earth's sediments point clearly to the involvement of comets fragmenting in the near-Earth environment over a 100,000 year interval. The main event that coincided with the extinction of the dinosaurs left its mark indelibly in the form of the Chicxulub crater in the Yucatan peninsula. Two other craters, one in Iowa and another in Kamensk (Russia) have also been associated with the mass extinctions of species that occurred at the K/T boundary. The evidence is that a giant comet plummeted into the inner solar system, passing close enough by Jupiter to fragment it into many pieces approximately 65.05 million years ago. Repeated passages by Jupiter over a 100,000 year period produced hierarchical fragmentation, and one such fragment (of normal comet size) came close enough to the Earth, perhaps first to become an Earth-comet, and then to crash onto the planet's surface.

Other mass extinction events at 36.9 million years, 94.5 million years, 200 million years and 248 million years BP have also been found to be associated with iridium enhancements in the corresponding sediments, so a cometary connection is believed to follow. Indeed the entire pattern of mass extinctions over the past 500 million years as seen in Figure E1 (from S.V.M. Clube and V.M. Napier, 1996) is strongly suggestive of recurrent catastrophic events of external origin with significant periodocities showing up in the record at 25 million years and 100-250 million years. Comet impacts are well suited to explain this chart. Sooner or later, within perhaps a few tens of millions of years another comet impact could cause mass extinctions as occurred 65 million years ago. Most life on the planet will then come to an abrupt end.

Besides causing mass extinctions on this scale, collisions with more modestly-sized cometary bodies and asteroids could have less

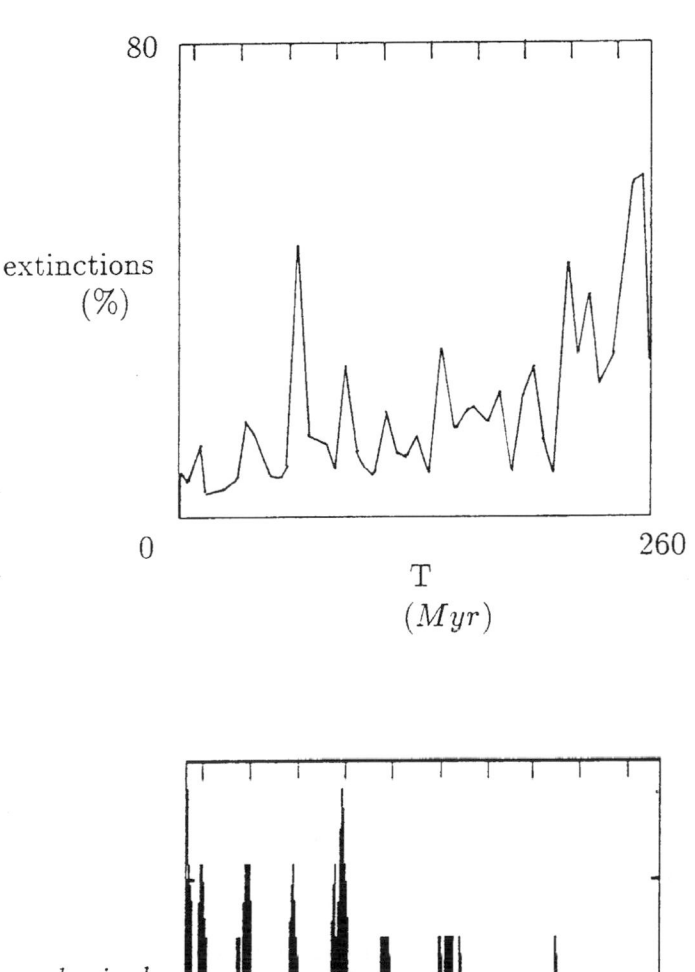

Fig. E1. Extinction record compared with major geological events attributed by Clube and Napier to comet impacts. (Courtesy of S.V.M. Clube and W. Napier).

dramatic, but nevertheless important results. About 20,000-30,000 years ago an asteroid with an estimated energy of about 15 mega-tonnes of TNT exploded over Arizona to form a 1.2 kilometre crater – the famous Meteor Crater. And a few other smaller crater sites have also survived over a few tens of thousands of years, all reminding us of an ongoing process and a continuing hazard. The history of human civilisation, if it is correctly read, bears witness to the most recent chapter in a series of cosmic events that controlled our planet in a decisive way. This chapter seems as if it began some 15000-16000 years ago with a giant comet of mass 10,000 million million tonnes becoming perturbed by Jupiter into an Earth-crossing orbit. (By Earth-crossing orbit is meant an orbit in which the comet is periodically brought closer to the sun than the Earth at its closest approach. This is a condition that is necessary if bits of the comet are to collide with the planet.)

Because comets in their early days contained radioactive heat sources, their interiors must have remained liquid for millions of years. When re-freezing occurred, starting from the exterior surface and freezing inwards, the comet would have developed cracks throughout and become considerably weakened in its structure The process is analogous to the cracking of pipes by freezing on cold winter nights. Successive passages of the comet by Jupiter would result in fracture and disintegration of the original giant comet into a multitude of smaller comets, perhaps eventually into a million separate pieces. These fragments would orbit the Sun on a potential collision course. For any particular fragment the chance of a colli-sion during a single orbit is small; with each orbital revolution the chance is only a few parts in a billion. Over a 15,000 year timespan, however, a few of the many pieces into which the comet divided would surely hit the planet to cause dramatic effects.

The first collision of this type seems to have taken place some 13000 years ago. At this time the planet was trapped in a highly stable ice age. The water that was released due to evaporation from the oceans probably restored the greenhouse effect on a very short timescale and caused the Earth to pass into a warmer phase. This is in accord with geological data that shows a warm pulse in the temperature profile nearly 13000 years ago. A further warm pulse, a second collision, followed close upon the heels of the first, 11000 years ago, and this event led to a decisive emergence of the Earth out of the last Ice Age.

The fragmentation of the original giant comet could not have stopped with sizes as large as a few kilometres (masses of a few hundred million tonnes). Further hierarchical fragmentation is bound to occur down to sizes of a few hundred metres or less. Because the smaller pieces are more numerous (in proportion approximately to the cube of the dimension) impact probabilities with the Earth will also rise accordingly. If a kilometre-sized comet hits the Earth every 10,000 years a 100 metre bolide will hit every 10 years and a 40 metre object will hit every year on the average. The effect of a comet or bolide impact on the Earth depends on several factors including mass, composition, structure, velocity, inclination of approach. For a low density cometary bolide of radius 40 metres hitting the Earth head-on at a minimum speed[1] of 13.5 km/s the kinetic energy of impact is equivalent to about 2 MT (megatonne) of the explosive TNT, that is, the equivalent of about a hundred bombs of the type that destroyed Hiroshima on August 6th 1945. Such an object, if made of ice, would mostly explode at about 30 km above the Earth. But an object of only 3 times this size would strike the Earth and cause destruction on a vast scale. A wide range of possible effects can arise with a bolide impact depending upon the missile energy as displayed in Table E1.

Although a detailed knowledge of the consequences of comet impacts does not exist at the moment, the general features of Table E1 cannot be too far off the mark. What remains to be discovered is how frequently these types of events have happened in the past, and how frequently might they recur in the future.

An object of about 100 metres across entered the upper atmosphere of the Earth over Siberia in the early hours of June 30th, 1908. A great fireball was seen to pass low over the town of Kirensk in Siberia and came down over a remote part of the Siberian forest.

---

[1]The collision speed of 13.5 km/s is appropriate for a comet with perihelion distance equal to the radius of Earth's orbit. For collisions with comets having appreciably smaller perihelion distances, the energies are about 10 times larger and the effects correspondingly more disastrous. The Table E1 is very much a minimum effects table.

TABLE E1

Energy	Diameter (for ice comet with v=13.5km/s)	Effects
1 MT TNT (50 Hiroshima bombs)	70m	Disintegration at high altitude, h>30km
10 MT TNT (500 Hiroshima bombs)	140m	Explosion in low atmosphere or surface impact; localised blast
10²MT TNT (5000 Hiroshimas)	322m	Localised blast damage; tsunamis
10³MT TNT (50,000 Hiroshimas)	700m	Extensive blast damage; fires, tsunamis
10⁵MT TNT (5 million Hiroshimas)	3.22km	Extended cratering; global effects; climatic changes; crop failures, famines
10⁶MT TNT (50 million Hiroshimas)	7km	Mass extinctions

The object did not reach the ground but exploded in the atmosphere at a height of about 8 kilometres. The brilliant fireball, said to outshine the Sun, was seen a 1000 km away from its point of descent, and the sound of the explosion was heard at even greater distances. The resulting blast wave felled trees over a distance of some 40 or 50 km, and the heat from the fireball charred tree trunks for distances of up to 15 km from the centre. This scene of devastation, first photographed in 1927, is shown in Plate 18. Estimates of the total energy of the impacting object range from 13-30 megatonnes of TNT, equivalent to the explosive power of 650-1500 Hiroshima bombs. The involvement of a cometary bolide rather than an asteroid seems likely in view of the observations of enormous and extensive clouds of reflecting dust particles (so-called noctilucent clouds) that were seen over a large part of the northern hemisphere. Furthermore, the estimated trajectory of the bolide points to a connection with the orbit of comet Encke. It is also noteworthy that the meteor stream associated with comet Encke crosses the Earth each year on June 30th, the precise date of the Tunguska impact. Again in 1965 on March 31st a brilliant fireball exploded over Revelstoke in Canada. It was heard and seen some 800 km away and tiny fragments of carbonaceous meteoritic material were recovered from the snow. The total energy released was estimated at about 20 kilotonnes TNT, equivalent to one Hiroshima bomb, but representing only a mini-Tunguska event it would seem.

Collisions of the Tunguska type, and others on a grander scale, must have occurred sporadically throughout our history and prehistory for the past 13000 years or more. At the beginning of this period, cometary activity, including dust production, following the break-up of the initial comet would have been conspicuously intense. If the cometary bolide that caused Tunguska was part of the stream created by the original giant comet this would probably have coincided with the brightest present-day meteor stream known as the Taurid-Arietid complex from which the Tunguska missile is thought to have come. When our ancestors encountered this stream 13000 years ago, twice yearly in June and November, the falling meteors would not have been as harmless as they are now. Indeed for much of the year the entire zodiac may have been seen to glow from the sunlight scattered by the newly evaporated cometary dust particles. Sightings of comets ferociously breaking up, and rising and setting with their long and graceful dust tails must have been exceedingly common in the ancient skies. Myth, legend and religion would undoubtedly have evolved in response to these remarkable experiences, experiences which must have been shared by many nomadic tribes that were scattered widely across our planet. Indeed major cometary fragmentation events would quite naturally have come to be immortalised by our ancestors into myths that depicted a struggle for supremacy amongst the gods that inhabited the heavens. Such is seen to be the case for most religions that have survived, and not unexpectedly we find a great uniformity of beliefs over widely different cultures of the world.

Scarcely a millennium after the end of the last Ice Age our ancestors discovered agriculture and farming, and thereafter began to relinquish their customary lifestyle as hunters and food-gatherers. As the first human settlements – villages and towns –

began to appear on the surface of our planet, assaults by cosmic missiles of the Tunguska – and super-Tunguska-type would seem to have occurred with unrelenting vigour. The kinds of effects listed in Table E1 would all have been experienced, and experienced repeatedly. Throughout our history the probabilities of cometary impact would by no means have remained uniform, as we shall shortly see. Rather one would expect to find significant bunchings of collisions, for the simple reason that fragmentation events due to planetary encounters occur on a discrete basis.

We can estimate that 7 major collision cycles occurred over the past 10,000 years and we shall attempt now to trace these events, worldwide, through recorded history. Table E2 lists the events we shall now proceed to explain.

TABLE E2
AN HISTORICAL CALENDAR WITH A 1,500 YEAR PERIOD

11,500BC	Giant comet breaks up near Jupiter
10,000BC	Collision of a comet: warm pulse
8,500BC	Second collision of comet: ice age ends
7,000BC	Bunched collision epoch
5,000BC	Bunched collision epoch Rice and millet cultivation in China
4,000BC	Bunched collisions Discovery of copper smelting
2,500BC	Bunched collisions Collapse of Mohenjo-daro End of Old Egyptian Kingdom Pyramid building
1,000BC	Bunched collisions Destruction of Jericho Homeric legends begin?
500AD	Bunched collisions Collapse of the Roman Empire Hun rebellion in India
2,000AD	New Era Begins??

The rises and falls of civilisations, the ascendancy and decline of empires, that punctuate human history over the last 10,000 years, can be explained elegantly on the basis of periodic or near-periodic assaults from the skies. The falls of major civilizations occur dramatically during the shorter bad periods and the ascents to power and glory would be sustained over the longer periods of calm. The bad periods quite naturally would have generated philosophies and even religions that were austere and harsh, while in the more extended quiet intervals mellowed and gentler attitudes prevailed. In this context it is important to note that the period of classical Greece, the Athens of Socrates and the India of Gautama Buddha and even the India of Asoka happened to coincide with extended periods of remission from cometary impacts. Similarly, the last five hundred years of our history in the run up to the year 2000AD, has also been free of major cosmic disasters.

The first crucial development in the chronicle of modern Man that may be related to comet impacts is the discovery of metal smelting. This was indeed a most remarkable discovery that led eventually to the use of metals for weapons and tools. It marked a turning point in the fortunes of man, the technological animal as he eventually turned out to be. The possibility of obtaining a shiny piece of metal from a stone could hardly have occurred to any human being as an abstract theoretical concept. So utterly impossible it would appear until after the process had actually been shown to work. One would have to see rock transmuted to metal before one's eyes to believe that it could ever happen. It stands to commonsense that the very first discovery must have been a case of serendipity – a fluke beyond a shadow of doubt. The problem, however, is to understand how the same remarkable accident could have taken place independently in all the widely-separated locations on the Earth where copper came to be used. Archaeological evidence shows that this metal was being used in making tools and utensils at a date somewhat before 4000BC. The relevant natural accident, that had to be repeated in several locations, was almost certainly the multiple impacts of cometary missiles. Events of the Tunguska-type could have started huge fires producing masses of glowing charcoal. Beneath the intense heat of glowing charcoal, rocks containing appropriate metallic ores would surely become smelted. Nomadic tribes chancing upon sites of smouldering fires similar to Tunguska 1908 would simply have picked up pieces of the smelted metal (copper, perhaps) and discovered that they could be beaten and flattened to yield artifacts that served their needs. The metal smelting epoch, as judged by archaeological evidence, does indeed tally with with a deep minimum in the averaged global temperature curve, thus strongly supporting a proposed link to comets.

The next period of intense bombardment seems to have occurred about 1500 years later at around 2500BC. Just before this time, several great civilizations are known to have flourished, both in ancient Egypt and in the Indus Valley of North India. The ruined city of Mohenjo-daro in northern Pakistan appears to have been the site of a pre-Aryan civilisation that was perhaps more advanced than that of ancient Egypt. It had flourished for over a millennium, but suddenly and dramatically collapsed. The fall of Mohenjo-daro from the greatest heights of glory presents a continuing puzzle to historians. Several possible causes have been discussed, but none has turned up that is entirely satisfactory. Aryan invasions from the West could have produced a slow erosion of the the empire, but not a seemingly cataclysmic fall. Likewise, seasonal flooding of the Indus valley could have had a slow cumulative effect, perhaps, but not a sudden one. A far more dramatic catastrophe may have arisen from tidal waves and tsunamis which arise quite naturally when cometary fragments crash into the sea.

The years around 2500BC also mark the abrupt end of the Egyptian Old Dynasty and the beginning of the Old Kingdom, the so called Pyramid Age. There are many intriguing mysteries that surround the building of the pyramids, an enterprise that has no parallel in the entire history of human civilization. Why did the Egyptians choose to build such stupendous structures that served no evident function except as royal tombs? Why did they scatter them over a vast expanse of desert, rather than build them all in one place? One thing at least seems certain. These structures were planned and built with meticulous care; they have survived over 5000 years and so must have witnessed several periods of Tunguska-type impacts. It is conceivable that the sizes and shapes of these structures were optimised for surviving the blast wave from a cometary object 100 metres in diameter exploding at a height of 5-10 kilometres above the ground.

Investigations of the three main pyramids of Giza have revealed some baffling alignments that strongly suggest an astronomical connotation. In the great pyramid of Giza there are two distinct channels that connect the main chamber to the outside world. Various interpretations have been offered for the existence of these channels – for example, ventilation, conveyors of food and provisions, and most interestingly perhaps, for astronomical observation of a specific kind. According to computations of Virginia Trimble some years ago one channel lines up with a bright star in Orion's belt at meridional crossing, and the other is in the direction of α-Draconis which was the pole star for the epoch 2600BC. Robert Bauval has recently argued that these orientations reflected beliefs of ancient Egyptians that the Gods lived in the heavens. The channels were intended to assist the dead Pharoah's soul to rise to a pre-determined point in the heavens. We ourselves would tend to favour the idea that the channels were used for astronomical observations. One could speculate that the directions of the shafts pointed to meridianal positions of the radiants of meteor streams that had recently led to destruction of the land. Another intriguing possibility is that the pyramids served the kings before their death as a kind of air raid shelter to protect them from cosmic missiles, and protect they must surely have done. Except for seemingly outlandish explanations of this sort the logic of the pyramids would remain an enigma – or if there was no logic they must surely represent a bizzare and incomprehensible aberration of the human intellect.

Keeping to a strict 1,500 year period, the next intense impact episode occurred around 1000 BC, which is a likely time of some of the events described in the Old Testament. Many of the Old Testament accounts of seemingly mysterious occurrences could have had a firm basis in fact if one admits the possibility of epochs of bunched cosmic collisions. Descriptions of deluge, a rain of fire on the cities of Sodom and Gomorrah, famines occasioned by the wrath of the gods – all have a rational basis as possible effects of cometary impacts. Fires, tsunamis or tidal waves, floods, climatic changes

adverse to crops, even clusters of earthquakes, can now be inter-
preted as real phenomena caused by the arrival of cometary missiles.
No metaphysical or mystical explanations are required. When
Joshua saw the Sun stand still in the sky, it might well have been
the glow of an immense fireball similar to what was seen over
Tunguska in June 1908. The two descriptions, in the Old
Testament and in Siberia of 1908, are similar.

At roughly the same time in China, that is to say, close to
1000BC, there were reports of 'widespread disturbances of the bar-
barians' according to bronze inscriptions of the Western Chou (c.
1122-771 BC) and these events evidently coincided with the fall of
Jericho and the beginnings of Judaism in the West. During this
period we are told that the authority of the Chinese Emperors
became reinforced, but only after diminishing for a while. A link
was then forged between the authority of the King and the power
of Heaven, and subsequently Kings were regarded as the Sons of
Heaven.

As pointed out by Victor Clube and Bill Napier in their excellent book *The Cosmic Winter*[1], the gods and goddesses of ancient
Greece were also located in the skies. We have noted that the break-
up of comets during collision episodes would have frequently led to
spectacular displays in the ancient skies, and such displays would
have come to be portrayed as wars between the gods. When any
prolonged episode of cometary break-up eventually came to an end,
the war of the gods would seem to have finished. But surely, on each
occasion, there would be a last highly conspicuous object in the sky
that would continue firing its jets of fire long after the others had
stopped. In Greek mythology, in the poems of Homer and Hesiod
(probably composed around the 8th century BC), this last object
may have become Zeus, King of the Gods, who with his bolts had
finally conquered the rest. It is likely that Greek mythology evolved
from the older mythologies of Western Asia and Mesopotamia that
date back somewhat before 1000 BC. If myths are to be based on
real events, it would seem reasonable to suppose that these events
took place at a time before 1000BC when, according to the

[1]*The Cosmic Winter* by Victor Cube and Bill Napier, Blackwell, Oxford, 1990.

calendar of events displayed in Table E2, the Earth had indeed suffered an episode of severe bombardment from the skies.

Even some 200-300 years after Homer and Hesiod, the Greece of classical times in the 5th century BC had not entirely forgotten the bolts from the blue, the disastrous events of bygone centuries. In the dialogues of Plato we thus find references such as the following in *Timaeus:*

> '*Critias*: . . . There were of old great and marvellous actions of the Athenians, which have passed into oblivion through time and the destruction of the human race ... There have been, and will be again, many destructions of mankind arising out of many causes; the greatest have been brought by the agencies of fire and water ... There is a story....that once upon a time Phaethon, the son of Helios, having yoked the steeds in his father's chariot, because he was not able to drive them in the path of his father, burnt up all that was upon the earth, and was himself destroyed by a thunderbolt. Now, this has the form of a myth, but really signifies a declination of the bodies moving around the earth and in the heavens, and a great conflagration of things upon the earth recurring at long intervals of time; when this happens, those who live upon the mountains and in dry and lofty places are more liable to destructions than those who dwell by rivers or on the seashore. When, on the other hand, the gods purge the earth with a deluge of water, among you, herdsmen and those of you who live in cities are carried by the rivers into the sea.
>
> . . . at the usual period (*crossing of the Tuarids?*), the stream from heaven descends like a pestilence, and leaves only those of you who are destitute of letters and education; and thus you have to begin all over again as children, and know nothing of what happened in ancient times. . ."

Dialogues of Plato (Timaeus) Translated by B. Jowett

It is clear from this that stories of comet and meteorite impacts were in currency right through to the time of Plato. The intellectual movement to forget and bury a turbulent past, and to deem the Earth safe from comets, seems to have begun in earnest with the philosopher Aristotle who lived in the period 384-322BC. With Aristotle, comets and meteors lost their exalted status as celestial objects, and were downgraded to become relatively minor effects in the atmosphere. The Earth became disconnected from the cosmos, at least in Western thought. These changes of attitude were made possible, scarcely 200 years after Socrates, for the simple reason that there then was a marked decline of all forms of meteor and cometary activity in the sky.

A mainly benign period started about 1000BC and lasted, with a few notable remissions, through classical times until the 6th century AD. Thereafter bad spells occurred intermittently through the dark ages, eventually returning to a benign period in mediaeval times.

The collapse of the Roman Empire in the 6th century AD has been the subject of intense scholarly debate for many years. Edward Gibbon, having first asserted on the authorities of Flamsteed, Cassini, Bernouli, Newton and Halley, that no harm could ever come from the skies, writes thus:

> '. . . history will distinguish . . . periods in which calamitous events have been rare or frequent and will observe that this fever of the Earth raged with uncommon violence during the reign of Justinian (AD527-565). Each year is marked by the repetition of earthquakes, of such duration that Constantinople has been shaken above forty days; of such an extent that the shock has been communicated to the whole surface of the globe, or at least of the Roman Empire. An impulse of vibratory motion was felt, enormous chasms were opened up, huge and heavy bodies were discharged into the air, the sea alternately advanced and retreated beyond its ordinary bounds, and a mountain was torn from Libanus and cast into the waves . . . Two hundred and fifty thousand people are said to have perished . . . at Antioch.'

> Edward Gibbon: *Decline and Fall of the Roman Empire,* Ch 43

The type of prolonged and frequent earthquake activity as described here is unusual and unknown in records of more recent times. An explanation has to be sought in terms of an external cause – cosmic missile impacts that could provide the trigger to send pressure waves into the Earth's crust and thereby generate prolonged bursts of seismic activity.

The collapse of Roman Britain in the middle of the 5th century AD was also connected with the same aggregation of cosmic missiles. There is a strong suggestion that Roman settlements in Britain were flattened through external causes as the following report in the London *Penny Magazine* (1834) indicates:

'. . . in the process of draining the Isle of Axholme in Lincolnshire, evidence has everywhere been found not only of previous vegetation but that this spot must have been suddenly overwhelmed by some violent convulsion of nature. Great numbers of oak, fir and other trees were lying 5 feet underground. It was reported that the tree trunks were all aligned north-west/south-east and had not been dissevered by the axe but had been burnt asunder near the ground, the ends still preserving a charred surface. . . .'

This description is strikingly reminiscent of the 1908 Tunguska scene that we have already described. The Isle of Axholme is a low-lying area some square kilometres in extent that was at the centre of a large settlement in Roman Britain.

Recently M.G.L. Baillie (*The Holocene*, 4, 212, 1994) obtained crucial evidence that supports this point of view. The evidence comes from the modern science of dendrochronology, which involves the dating of events by essentially counting the numbers of tree rings and measuring their thicknesses. Every passing year yields an additional ring in the trunks of trees, and the thickness of tree rings in a given year can be used as a measure of the amount of sunlight during the growing season. If for any reason sunlight is diminished, the tree-rings corresponding to those years are also correspondingly thinner. Baillie's studies of tree-ring thicknesses, corresponding to the early decades of the 6th century AD, reveal that there is a major dip in the Earth's temperature over the entire period AD536-546. The idea that a volcanic eruption was responsible for a dust shroud that lowered the temperature and reduced seasonal tree growth does not square up with the lack of an acid signal in Greenland ice-drills of this same period. Furthermore, volcanic dust is known to settle in a couple of years at most, and cannot therefore explain such a protracted episode of depressed average temperatures.

There is literary evidence as well: contemporary writings comment that 'the sun was dark and its darkness lasted 18 months . . . the sun appears to have lost its wonted light and appears of a bluish colour . . . fruits did not ripen . . . cold and drought finally succeeded in killing off the crops in Italy and Mesopotamia and led to terrible famine in the immediately following years' . . . and so on.

Intense meteor and cometary activity has been recorded even in the closing decade of the 6th century AD. For instance in an Islamic text cited by the late W.M Smart, one finds the following graphic description:

> 'In the year 599 on the last day of Moharrem, stars shot hither and thither and flew against each other like a swarm of locusts; people were thrown into consternation and made supplication to the Most High'.

In India it would also appear that the dawn of the 6th century AD marked a time of exceptional turbulence and strife. In J.C. Powell-Price's *A History of India* we are told that:

> 'The eruption of the Huns had a very great effect on India. It broke up the empire of the Guptas and upset the developments which the Gupta culture was leading to. These barbarous hordes were out for plunder, blood and rape and had no feelings for either art, literature or religion. . . . The history of N. India is very confused after the eruption of Huns in about 500AD.'

To recount the overall course of events that followed the break-up of our 'Civilization Comet', reference to Table E2 shows that the major events in our history that can be easily fitted to a 1,500 year collision period. For at least 4 cycles in the run-up to 2000AD the correspondences seem to be well-attested. With the possible exception of a brief period around the year 1200AD, we note that there is no evidence of major incursions of cometary fragments from the middle of the 6th century AD all the way to the present-day. On the basis of this calendar, showing repetitions over a period of 1,500 years, the next major events are to be expected in the decades, perhaps century, following the year 2000AD. One might hope that this time round, the meteor stream associated with our Civilization Comet has become so attenuated and so denuded that the forthcoming period of bunched impacts may not lead to catastrophic effects, at any rate not on the scale that has occurred before. However, as we shall see, it is as well to be forewarned and hopefully to be prepared.

During the lengthy period of relative safety that followed the collapse of Rome, it would seem natural for humans to invent a social philosophy that devalued or ignored the effect of events in the sky. The Church of Rome, already at the end of the 7th Century, had decreed that Man was to be disconnected from the Cosmos. The heavens were unquestionably benevolent; the stars were studded in the celestial vault merely to serve as an adornment and to extol the magnificence of God. Post-Newtonian science provided the cultural framework reaffirming the Church's edict and for confirming the older Aristotelian idea of a Universe that did not, and could not, impinge on the affairs of men. By the closing decades of the 17th century AD it may well have been seen that a benevolent cosmos governed by Newton's laws was the ultimate triumph of the Aristotelian world view. According to Victor Clube, Newton is said to have initially abhorred the idea of comets in Earth-crossing orbits, because he clearly foresaw the horrific prospect of collisions with comets if they were solid bodies. Scientists, however, quickly disposed of this fear by asserting without proof that comets are insubstantial tenuous objects. With the growth of Newtonian science one could say that so-called uniformitarian traditions in science came to be established. The whole of terrestrial evolution including the evolution of life had to be accommodated thereafter within an intellectual framework in which no catastrophes, no sudden changes, no intervention of the heavens could be allowed.

Scientific academies and establishments throughout Europe conspired to maintain this position late into the 18th and 19th centuries, and indeed aspects of it survive to the present-day. For a long time, during the seventeenth and eighteenth centuries, it was maintained that meteorites (stones of extraterrestrial origin) could not even exist. Whenever evidence to the contrary was presented, 'learned' societies such as the French Academie de Sciences vehemently denied the evidence, maintaining that stones of extraterrestrial origin simply did not exist. It was only when a fall of over 2000 'stones from heaven' fell at L'Aigle on 26 April 1803 that the Academy changed its mind, and then only because it was impossible to deny such a prodigious fall seen by a multitude of people. Present-day attitudes to even the strongest evidence of cosmic life, such as were discussed earlier in this book, bear the

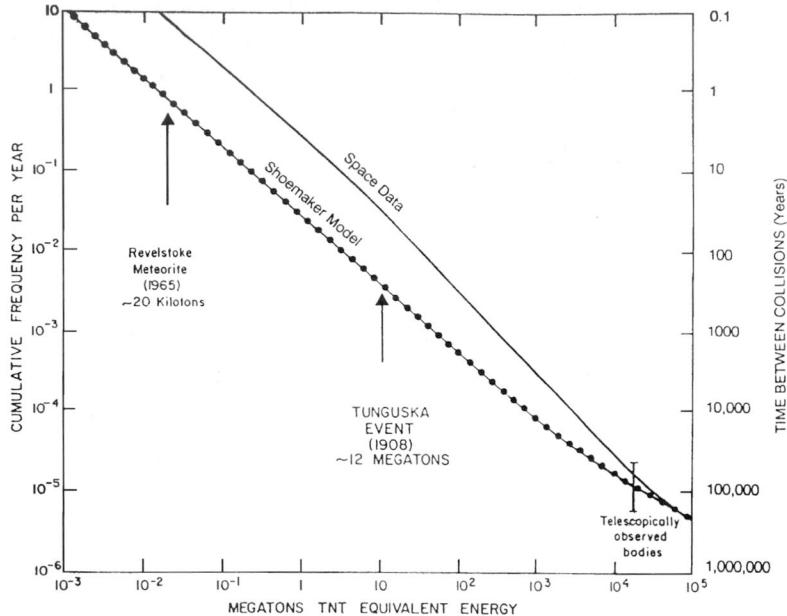

Fig. E2. Cumulative frequency of impacts up to various energy limits.

hallmark of the same deep-rooted prejudice. Life, according to current dogma, is a purely terrestrial phenomenon, and the wider universe must be kept well clear of its purview. Such prejudice can be
seen as the legacy of the old Aristotelean view of the world.

In conclusion we turn to modern scientific attitudes and some
modern results. Recent studies of craters on the lunar surface have
given important clues that could relate to impacts on the Earth.
The line of connected points in Figure E2 shows the frequency of
impacts on the Earth estimated from models using information
about the size distributions of the youngest craters on the Moon.
From this curve (due to Prof. E.M. Shoemaker) it would seem that
the occurrence of a Tunguska type collision can be expected perhaps
once per century. Surveys of the Earth from satellites have, however,
revealed that sub-Tunguska missiles are arriving globally at an
alarming rate, a rate that would seem to demand that the dotted
curve in Figure E2 has to be lifted *at least by* a decade. Between
1975 and 1992 some 136 NEO's (near-Earth objects) impacted the
Earth and were detected by space-based infrared and visual sensors.

None reached the Earth, nor came near enough for any of the adverse effects of Table 1 to be realised. However, the energy estimates of the observed events which were in the MT (TNT) range (50-100 equivalent Hiroshima bombs) suggest that a large fraction were only marginally below a critical value for exploding close to the surface and causing local devastation.

This raises a question of pressing urgency as to when and where the next Tunguska-sized or larger cometary bolide might strike. Earlier estimates of one per century may need to be revised upwards to perhaps once per decade. Also there is the possibility that we may soon be approaching a 'bunched period' of impacts, in which case the situation would be more alarming. Scientists are at last recognising that it would be prudent to start an observational program that would search the Earth's immediate space environment for Near Earth Objects (NEO's) that could possibly be heading towards our planet. What seems to be required is a world-wide network of small telescopes dedicated to such a project, and the development of deflection and mitigation strategies in the event of discovering an object that threatens us in the future. At a recent international workshop held at Erice, Sicily (April 28-March 4, 1993) the following statement was signed by some 60 delegates:

> 'The gathering of additional data on Near Earth Objects and their effect on Earth is a scientifically and socially important endeavour. These efforts should be conducted in a coordinated international manner. Dedicated international facilities similar to the proposesd Spaceguard System should be developed. The defense related assets and technologies of the former Cold War combatants can contribute to the gathering of these data, through ground and robotic space observation. These skills and technologies, necessary for any large complex investigation, should be well utilized now that the threat of global thermonuclear war has been reduced.....'

At the present time several international programs are underway to search the skies for NEO's, but the total manpower and financial resources devoted to this venture remains small. The surveys in progress include the Palomar Asteroid and Comet Survey (PACS), which deploys the 0.46 metre Schmidt telescope on Mt. Palomar, the University of Arizona 'Spacewatch' program, and the

Anglo-Australian Near-Earth Asteroid Survey (AANEAS). The current rate of discovering new NEO's amounts to about 20 per year, but for most of these there is not enough data to determine accurate orbits. It is necessary to intensify our efforts to establish a more complete database of NEO observations from which secure predictions might be made. There are hopeful signs that the scientific community is ready to countenance such a scheme. Clearly more effort is needed, and there is every sign that this will come.

Mankind may at last be waking up from an amnesic slumber that has lasted for over 500 years.

APPENDIX

A more sophisticated theory of influenza

Introduction

Suppose a well-defined laboratory strain of influenza virus, V1 say, is used to infect an experimental animal. Let the output of virus, V2 say, be used to infect a second animal, and let the output of virus from the second animal, V3 say, be used to infect a third animal .... and so on. thereby generating a sequence of case-to-case transmissions under controlled laboratory conditions. It is found in such a series that:

$$V1 \cong V2 \cong V3 \cong \ldots \cong Vn.$$

The final output virus Vn is essentially similar to the initial input virus V1. Yet, if such a series is generated with the first case coming from a virulent natural infection, it is found that Vn lacks the virulence of the initial infection.

It has been usual to suppose that for a series initiated by a natural infection, the virus changes in some way, perhaps due to copying errors along the chain, but this is contrary to other evidence which shows that genetic material is copied in cells with very high efficiency, as for instance we have discussed in Chapter 7. It is also contrary to what is believed to obtain in pandemic situations where Vn is supposed to maintain its virulence even though the sequence of case-to-case transmissions is very long.

One way to escape from this paradox is to drop the implicit assumption that natural influenza is caused by a fully-fledged virus. To examine this possibility, we choose a more general model,

$$A1 + A2 + \ldots \to \text{Animal} \to \text{Virus},$$

where A1, A2, ... are activation factors that need not be fully-fledged virus. Some of the factors are to be regarded as essential if the disease occurs, but others may be optional.

As a particular case of this general model we discuss the possibility:

$$A1 + A2 + A3 \rightarrow \text{Animal} \rightarrow \text{Virus},$$

where A1 is a triggering agency (a viroid), with the genes of the output virus being contained in either A2 or A3. The reason why A2 and A3 are presented as alternative sources of the viral genes will appear at a later stage. The first essential is to convince oneself of the need for the trigger A1 and that A1 must be incident on the Earth from space.

Reasons why there must be at least one worldwide activation factor

Before Pasteur's work showed microbes to be the agents whereby diseases and other forms of biochemical activity are transmitted from place to place, the strange epidemiology of influenza had suggested that conditions in the Earth's atmosphere controlled the incidence of the disease. In his book *The Modern Practice of Physic*, published in 1813, Robert Thomas commented:

> 'By some physicians influenza was supposed to be contagious; by others not so; indeed, its wide and rapid spread made many suspect some more generally prevailing cause in the atmosphere.'

Pasteur's work demonstrated in the opinion of most doctors and biologists that influenza had to spread by contagion. Although the facts of epidemiology remained just as cogently against this opinion as they had been in the time of Robert Thomas, consciences were salved by the suggestion that, if one looked for long enough in the opposite direction, the facts would somehow go away. They have not done so of course.

On the contrary, a large body of data has accumulated over the years that decisively goes against the doctrine of contagion. Some of this data has already been discussed in Chapter 6. We shall now consider further evidence for the occurrence of *at least* an activation factor for influenza on a global scale.

We already showed in Figure 6.6 using data obtained by R.E. Hope-Simpson (*J. Hyg. Camb.*, 83 (1979), 11) for families in a general practice that there was little, if any, evidence of transmission under living conditions of the greatest personal intimacy. We ourselves obtained similar results for the outbreaks of influenza that occurred in English and Welsh boarding schools during the late winter of 1977-78. Although the main feature of our survey was its statistical weight, there were about 10,000 victims in a total school population of about 25,000, in a few schools we were able to obtain precise details of the positions of the beds which had been occupied by victims in the dormitories. The situation for possible transmission was then similar in principle to the households of Hope-Simpson. Like him, we found no evidence that transmission had taken place.

The surprising new feature of our survey was the extreme variability in the incidence of cases according to the physical position occupied by victims within their respective schools. We discussed our findings for particular schools in Chapter 6. Although pupils in the various school houses met together for instruction, for games, and in some cases for meals, there were nevertheless enormous fluctuations between school houses that were quite beyond what could he attributed to chance.

The facts do not permit any other conclusion than that human activities have little to do with the incidence of influenza. Pandemics sweep over the Earth in their own good time, unhurried by the intensity of human travel or personal contact. The primary agent of the disease evidently lies outside humans, and its attack when examined in detail is exceedingly capricious.

The local details of epidemic attacks on populations distributed over areas with dimensions of a few tens of yards on the low side up to a few hundred miles on the high side accord very well with the meteorology of a storm pattern.

As we have shown elsewhere (*Diseases from Space*, Dent, London 1979), the picture is of the primary activating agent being carried downward in water droplets from the tropopause, of the water droplets evaporating before reaching the ground, and of the causative agent of the disease then blowing about the lower atmosphere. To quote from our former discussion:

'Strong winds favour the generation of turbulence, although on their own they are by no means sufficient. A wind can blow over a smooth level plane or over the ocean without the generation of turbulence, but a wind blowing over an irregular surface becomes turbulent if it is strong enough. For wind blowing over the land there are many scales to the turbulence. The largest scale would be induced by major obstacles such as hills and mountains, which would give rise to eddy interchanges up to thousands of feet and would thus be responsible for bringing pathogenic particles (now freed of their raindrops) down to within hundreds of feet of the ground surface. Local irregularities, such as buildings and woodlands, generate intense fine-scale turbulence which can then reach upward for these last remaining hundreds of feet. Local turbulent eddies generated by minor irregularities are, therefore, finally responsible for the detailed incidence of the pathogen at ground level. There may be fortunate spots which are either smooth enough or sheltered enough not to send local eddies upwards through these last remaining hundred feet or so. For these spots, the pathogenic particles will fly by in the wind, immediately above the heads of people who fail to realise their good fortune.'

Reasons why the primary activation agent must be incident from space

The worldwide spread of influenza pandemics cannot be caused by storm patterns, however. Storms are local distributors, whereas the broad global spread of influenza is a larger-scale phenomenon. The distinction corresponds very well to the difference between the lower troposphere of the atmosphere and the higher stratosphere. The tropopause, which separates the lower from the upper regions of the atmosphere, is at a height of about 12 kilometres over middle latitudes and at about 18kilometres over lower latitudes. It is a crucial property of the atmosphere that up to the tropopause the usual situation is for the air temperature to fall with increasing height, to a value of about $-60°C$ ($213°K$), as shown in Fig. A.1. With increasing height from the tropopause up to about 20 kilometre altitude, the temperature remains close to this value, and then increases steadily to a maximum near $0°C$ ($273°K$) at an altitude of about 50 kilometre. This region of increasing temperature is the stratosphere, and the altitude of the temperature maximum is the stratopause. Above the stratopause comes the mesophere, which like the lower troposphere is a region in which the temperature decreases with increasing altitude.

The behaviour of the temperature with altitude is relevant to the influenza problem because vertical atmospheric motions of a

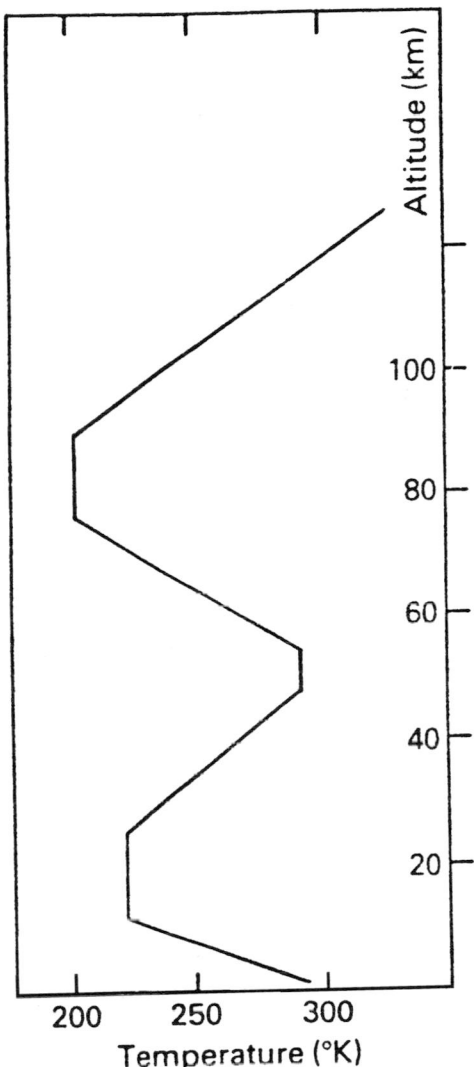

Fig. A.1. The variation of temperature with height in the atmosphere.

local kind occur easily when the temperature decreases with altitude (i.e. in the troposphere and mesosphere) but local vertical
motions do not occur in the stratosphere. It is only because the
atmosphere is heated non-uniformly and because the Earth is spin-
ning that any vertical motions occur at all in the stratosphere, *and
such motions as do occur are global, not local.*

Consider now the small particles from cometary sources which are entering the Earth's atmosphere at a rate which in aggregate amounts to about 100 tonnes per day. Particles of sizes of 1 μm or less do *not* burn up as they enter the atmosphere. They land 'soft' in the atmosphere at altitudes from 110 kilometres upward. If there are bacteria, viruses and viroids in the cometary material, they can land safely on the Earth, which they could not do on a body without an atmosphere like the Moon.

The atmospheric gases at such high altitudes are very diffuse, and the cometary particles fall under Earth's gravity below the temperature minimum at the 90 kilometres level in a matter of hours. In the mesosphere they are also soon carried down to the temperature maximum at the stratopause. This happens because the gases in the mesosphere are constantly in vertical motion.

Further descent from 50 kilometres to 20 kilometres is a different story, however, because the atmospheric density is now much increased so that the time of descent under Earth's gravity is changed from hours to years. Because there is no assistance from local vertical air movements in this region, the stratosphere therefore becomes a trap for the particles.

There are three ways for particles to escape from the stratospheric trap and to come down to the troposphere:

(1) Particles with larger sizes of the order of a micrometre are pulled down to the troposphere in two or three years by Earth's gravity. Viruses or viroids would need either to be clustered or to be encased in a matrix of protective material to take advantage of this mode of descent. In a naked form their smaller sizes would make their times of fall so long, many years, that other modes of descent would be more important.

(2) If a particle contains substances that emit photoelectrons the particle will become positively charged. Quite large electrical fields are generated from time to time in the stratosphere, for example through outbursts of particles from the Sun impinging on the Earth's magnetic field, and through powerful thunderstorms in the troposphere extending their electrical effects into the higher

atmosphere. Charged particles, if they are small enough, can be forced downward against the frictional resistance of the atmospheric gases by such fields. In this case it is the smallest particles, as for instance naked small viruses or viroids, for which the effect is strongest. Unlike the other two modes of descent, which are broadly global, involving time-scales of a year or years, this mode is essentially local and instantaneous.

(3) Each year, mostly between latitudes 55° and 65°, there is vertical mixing of the stratospheric air. In the northern hemisphere the mixing begins in early November and extends through January. In the southern hemisphere the same phenomenon occurs but with a time displacement of six months. The mixing extends the whole way from the stratopause down to regions immediately above the tropopause (from which a few weeks journey brings the particles down to ground-level).

If we restrict ourselves to particles of the sizes of viruses and viroids, modes (2) and (3) are of main interest. The capricious strikes of mode (2) are in good correspondence with the initiation of the second wave of the 1918 pandemic, while mode (3) corresponds closely to the well-known annual influenza 'season'. As regards mode (2), it is worth noting that outbursts from the Sun are strongly correlated with the 11-year sunspot cycle. In this connection it has been pointed out by Hope-Simpson (*Nature*, 275 (1978), 86) that the pandemics of the twentieth century have tended to break-out at the times of sunspot maxima, which are just the times when solar outbursts tend to be strongest. In *Diseases from Space* (p.175) we examined the relation of pandemics to sunspots for the eighteenth and nineteenth centuries, finding that Hope-Simpson's association continued to hold good generally, although not quite so precisely as in the present century.

As far as we are aware no sensible suggestion other than mode (3) has been put forward to explain the annual winter influenza season. The alternation six-months apart of winter seasons of the disease in the northern and southern hemisphere is a sure indication of a phenomenon involving the whole of the Earth's atmosphere.

Ample evidence of such a six-months displacement is available from influenza records in Australia and S. Africa for the south and from the United States and Europe in the north (See also Fig. 6.13).

The equator is in a neutral position between the geographical hemispheres. What happens then as the equator is approached from both the north and the south? This question has been partially answered by Hope-Simpson with the aid of data from southern, central and northern Africa for the period May 1950 to April 1951. From Fig. A.2 it is apparent that outside the equatorial band from about 15°S to 15°N (contained in the middle section marked 'tropical' in the figure) there is a clear phase lag of about six months between the two hemispheres. Within the 'tropics' thus defined there is a bias towards the south, but with an indication of a transition zone from French West Africa (Senegal, Ivory Coast, Dahomey, at latitudes of about 10°N). Likely enough, the transition zones are not the same at all longitudes, being biased sometimes south and sometimes north of the equator. Unfortunately, data for latitudes 20 to 25°N are missing from Fig A.2, because the lands in these latitudes are desert with low scattered populations. However, the broad features of the transition across the equator can clearly be seen. (See also Fig. 6.13).

Good data for the early winter breakthrough at middle to high latitudes in the stratosphere were obtained from the radioactive isotope rhodium-102 generated in the HARDTACK atmospheric nuclear bomb test of 11 August 1958. The explosion occurred at about 43 kilometres altitude above Johnston Island (16°N, 170°W). The nuclear debris from the explosion went overwhelmingly upwards to heights above 100 kilometres, where the radioactive material spread quickly around the whole world. Because the Rh-102 was not particulate, the individual atoms took a time of the order of a year to appear in quantity at the stratopause – unlike even small particles like viroids, the Rh-102 atoms were held up for a considerable time in the thin stable air above 90 kilometres. When by mid-1959 they reached the stratopause, however, the atoms of Rh-102 provided a highly effective tracer material for determining global motions downwards through the stratosphere. Data were obtained from samples of air taken at an altitude of

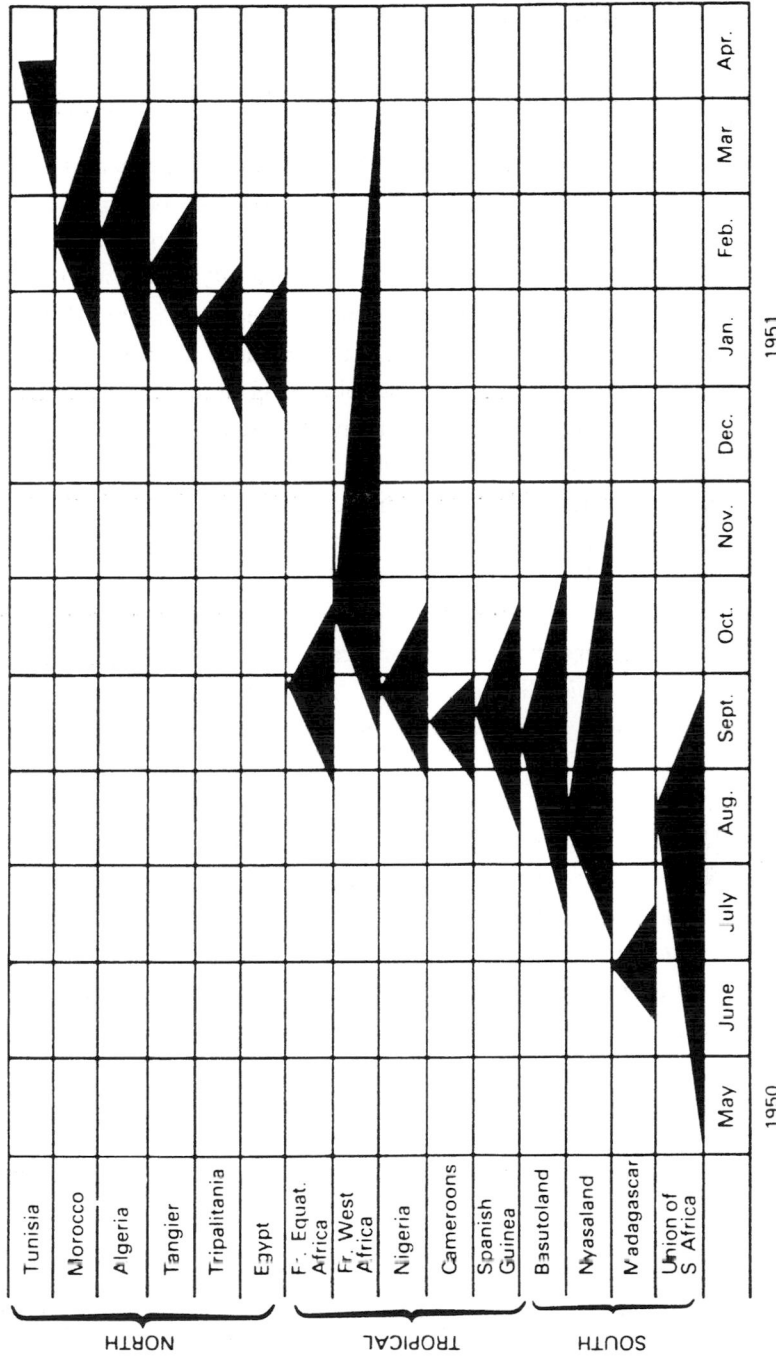

Fig. A2. Hope-Simpson's data illustrating the incidence of an influenza epidemic as a function of geographical latitude.

about 20 kilometres, which is close to the bottom of the stratospheric trap. From altitude 20 kilometres down to ground-level would be a matter of a further two or three months for individual atoms, but perhaps only two or three weeks for particles. The samples were taken by plane and balloon at various latitudes and longitudes, results being reported by M.I. Kalkstein (*Science*, 137 (1962), 645).

Figure 6.11 (to which we already referred briefly in Chapter 6) shows the rising concentrations of Rh-102 that occurred in the northern winter of 1959-60. The primary break down through the stratosphere occurred near latitude 65°N. This was followed in November to January by a major break at 45 to 50°N. (The smaller effect at 15-30°N is not stratospheric, however. It is to be attributed to a horizontal spread of Rh-102 at altitude 20 kilometres taking place from higher to lower latitudes.)

It was possible to estimate the total amount of Rh-102 produced in the nuclear explosion. Comparing the estimated total with the quantities recovered in the samples it was found that it would take of the order of a decade for the whole of the Rh-102 injected into the stratosphere to reach ground-level. This is strikingly similar to the time interval between major influenza pandemics. However, it should be noted that small particles at the stratopause over the extreme polar regions would probably require several decades to clear themselves entirely. The polar caps are in the nature of regions of longer term accumulation.

The epidemiologic data for influenza forces one to take the stratopause as the place of supply to ground-level of the primary causative agent, denoted above by A1. Because small particles from space arrive inevitably at the stratopause, whereas particles from ground-level would have to fight both Earth's gravity and the general stability of the air from 20 kilometres altitude up to 50 kilometres, it is natural to think of A1 as being space-incident. The only exception we have been able to think of to this argument would be an ejection of A1 in the outbursts of volcanos. Although some might prefer a volcanic source to incidence from space, the idea is ruled out by the need of A1 to have survived in that case for long periods inside the Earth at temperatures in excess of 1000°C.

Even if we ignore the physical difficulty of A1 climbing from ground-level up through the stratosphere, a source of A1 at the ground would long ago have been noticed by epidemiologists.

Specification of a model and the initiation of the disease

We return now to the model,

$$A1 + A2 + A3 \rightarrow \text{Animal} \rightarrow \text{Virus},$$

specifying A1, A2, A3 as follows:

A1 is a space-incident viroid or genetic fragment capable of 'rescuing' inactivated virus[1] which has already been added to the respiratory cells of an animal. If A1 enters a cell and finds no inactivated virus there, it is quickly destroyed by the enzymic apparatus of the cell.

A2 is a space-incident virus, or portion of a virus, added inactively to the genome of respiratory cells.

A3 is an inactive virus of terrestrial origin added to the genome of respiratory cells.

It is encouraging that many consequences and insights follow easily from this specification of the model, which requires that influenza can arise only if A1, enters a cell where *either* A2 or A3 is present already.

Clean Communities

A clean community is one without A3, without horizontally-transmitted inactive virus. No person in such a community can contract influenza unless A1 and A2 happen to come together in the same respiratory cell. For any one person this would be a highly unlikely event, because with both A1 and A2 space-incident and breathed by each person only in small numbers, it is improbable

[1]For an example of a 'rescuing' operation, see M. Park, D.M. Lonsdale, M.C. Timbury, J.H. Subak-Sharpe and J.C.M. Macnab, *Nature,* 285 (1980), 417.

that they will ever come together in the same cell so as to initiate the disease.

Since A1 is quickly destroyed if it does not find a 'mate', A1 does not accumulate in the respiratory cells. But the inactivated virus A2 accumulates, increasing the chance of the disease being contracted. This tendency is limited, however, by the natural death of cells and by their replacement with 'cleans' cells. For a solitary individual, never exposed to A3, such clean replacements of dead cells would keep the probability of contracting the disease very small throughout a whole lifetime. For a community, the probability of a single outbreak of the disease multiples with the size of the population, but if the population is small, limited say to only a few hundred persons, the situation remains that nobody is likely to experience the improbability of A1 and A2 coming together in the same cell. Thus a small isolated clean community tends to remain clean.

Dirty Communities

A dirty community is one with A3. The presence of A3 implies that someone in the community has experienced the disease. Other members of the community have breathed virus, far larger in quantity than A2, that has been broadcast by the victim. Because such transmitted virus is hardly ever active, the disease itself was apparently not directly transmitted, in accordance with the epidemiological facts discussed earlier. Yet those who have been in contact with the victim have now had large numbers of their cells 'primed' by A3, and will, therefore, be exposed with far higher probability than in a clean community to themselves succumbing once A1 comes along again.

The route is evidently opened to an escalating situation. If the first victim spreads sufficient A3 to infect at least one other person, the virus A3 spreads eventually throughout the whole community. The community becomes dirty, and likely enough it remains dirty, since once there are many victims the amount of A3 that is spread around is so large that, short of a decisive immunity developing, the disease becomes self-perpetuating.

All very large communities must be dirty. Even if we suppose an initially clean situation, with each individual having only a tiny probability of A1 and A2 coming together to set off the disease, for a large number of individuals the sum of sufficiently many tiny probabilities will add to unity, and an outbreak must happen. Then A3 will begin to spread progressively and within months influenza will explode throughout the community, even though there has been no direct case-to-case transmission of the disease.

All young babies are clean, at any rate in the present sense. They are not seriously exposed initially to risk of infection by influenza. Like individuals in a clean community, there is little chance of A1, and A2 coming together in their respiratory cells. And because young babies do not usually spend much time in public places, it may be a considerable while, even in a dirty community, before they acquire A3. The initial cleanliness of somatic cells at birth could be a major factor, both for influenza and for some other diseases, in the seeming immunity of babies. Sooner or later, however, a fond parent or relative goggling over the child passes on A3, apparently without infection taking place. The child is then primed for its first attack of the disease which will inevitably be a bad one because no actual immunity yet exists. This leads to an immense jolt on the immune system, a jolt which has been described as 'original antigenic sin', a jolt which gives a lifelong bias to the system.

In this connection we note a curious situation that came to our notice. During the period 15-20 February 1981 a wave of what seemed like the common cold ripped through a maternity ward at the University Hospital of Wales in Cardiff. Almost all the mothers and nursing staff were affected but remarkably the disease did not seem to pass to the new-born infants, indicating that the new-born infants were not yet primed for attack by the space-incident respiratory virus.

It is inevitable that cometary sources of A1 must be irregular in their supply to the atmosphere. Because it takes several decades for the whole atmosphere (including the polar caps) to clear itself of any particular injection of A1, irregularities of supply on time-scales of a few years are smoothed in their incidence at ground-level. If, however, the supply were interrupted on a time-scale of half a century, with the atmosphere being given time to clear itself entirely of A1, all communities, small and large, would become clean. This would happen through older persons initially with A3 gradually replacing their respiratory cells, as well as through new births.

The subsequent resumption of A1 and A2, contingent on a new supply from a cometary source into the atmosphere, would then lead, after a slow beginning, to devastatingly explosive outbreaks, especially in densely populated areas, with weakly developed or non-existing immunities causing high mortality rates. Such seems to have been the situation in 1889-91 as shown in Fig. A.3.

A confirmation of the model from a remote geographical source

It was only after we had arrived at the above model that we recalled, from a distant corner of the memory, a paper which had seemed quite inexplicable when we first read it. As one always tends to do with things that seem inexplicable, we had conveniently forgotten about it in the interim.

Before the clearing of the tropical forests of Brazil and Surinam, tribes of Indians lived alone in isolation, typically with populations of only about 500 persons. With the clearing of the forests these tribes were contacted and blood sera were obtained from some of them. From antibody tests, both 'clean' and 'dirty' communities (in the sense we have defined earlier) were found. An example of a 'clean' community was reported by J.V. Neil, F.M. Salzano and P.C. Jacqueira (*Studies on the Xavante Indians of the Brazilian Mato Grosso. Hum. Cenet.*, 16 (1964), 52). Similar results were obtained by F.L. Black *et al.* (*Am. J. of Epidemiol*, 100 (1974), 230).

The need for A2

Although A1 and A2 can find each other only very rarely, it would be incorrect to suppose that A2 plays no role except as a start-up factor in converting an initially clean large community into a dirty

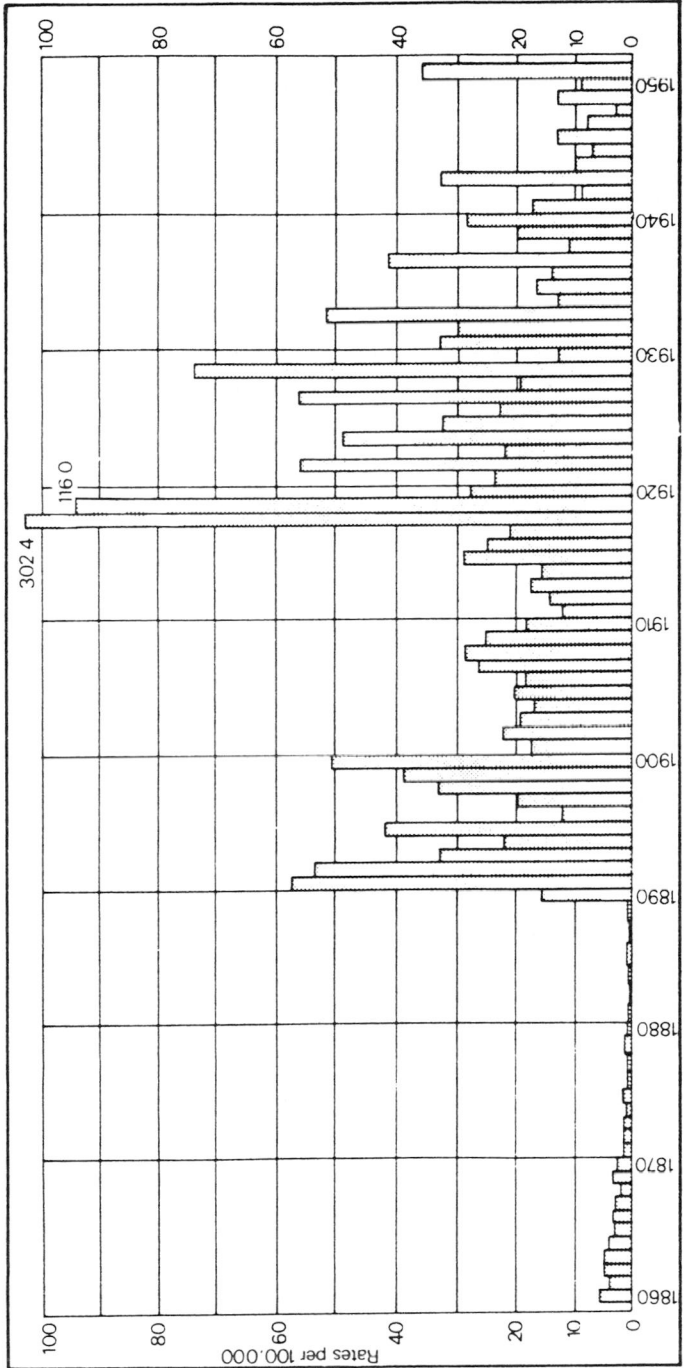

Data Based Partly on the League of Nations Health Organisation's Annual Epidemiological Reports

Fig. A3. Annual death rate from influenza in England and Wales.

one, as may have happened in 1889-91. In cases where A1 and A3 set off an attack of the disease, virus particles spread in large numbers throughout the respiratory cells. It then becomes probable that a cell containing A2 will be invaded by an active particle generated from A1 and A3. This may have the effect of 'rescuing' A2, or of promoting a recombinant event in which a new viable particle is generated partly from A2 and from the progeny of A1 and A3.

If other things were equal from an immunological point of view, such a new particle would be swamped in number by the progeny of A1 and A3, and would therefore be of little relevance to the clinical aspects of the disease. But if A3 persists in a community for a period of years, group immunity will be built up to the point at which few clinical outbreaks occur, although sub-clinical attacks must continue in order that the immunity process be set in operation. Sub-clinical attacks can serve to 'find' A2, and if either the rescued A2 or a recombinant virus involving A2 happens to outflank the immune system it will be the resulting new viral particle that will persist and grow in number, beyond the stage when the progeny of A1 and A3 are brought to a halt in their growth by the immune system. Thus a new antigenic type of virus will emerge at the clinical level. What happens then is that the new antigenic type replaces A3 in the community. The disease shifts to a new form which continues until improved immunity to it, together with the arrival of a further A2, repeats the process.

Work on the H1N1, H2N2 and H3N3 influenza subtypes by C. Scholtissek, W. Rhode, V. von Hoyningen and R. Rott (*Virology*, 87 (1978), 13) strongly suggests that the Asian flu virus H2N2 emerged in 1957 as a recombinant generated from H1N1 (the previous A3) and some A2, while H3N2 emerged in 1968 as a recombinant derived from H2N2 and some other A2. Thus A3 was represented in turn by H1N1 (up to 1957), by H2N2 (1957 to 1968) and by H3N2 (1968 to 1978).

The re-emergence of H1N1 in 1977-78 cannot be explained as a recombinant event, however, since the H1N1 of 1977-78 was nearly identical with that of 1950.[1] This seems to have been a case

---

[1]K. Nakajima. U. Desselberger and P. Palese, *Nature*, 274 (1978), 334.

in which A2 (H1N1) was rescued substantially without modification, and in which it competed in 1977-78 on about equal terms with the previous A3 (H3N2). The result was that for a while A3 had a dual identity. This led in 1978-79 to the detection of recombinants between the two forms of A3 (J.F. Young and P. Palese, *Proc. Natl. A cad. Sci. USA*, 76 (1979), 6547; A.P. Kendal et al., *Am. J. Epidemiol*, 110 (1979), 462).

The need for A2 to be space-incident

Through the importance attached to the horizontally-transmitted A3, the present model has come some way towards the usual theory of influenza transmission. The approach towards the usual theory is more apparent than real, however, because A3, although horizontally transmitted, is considered to be derived from A1, and from a previous A2, both of which are space-incident. In effect, therefore, A3 is also space-incident.

If it could be argued successfully that A2 had a terrestrial origin, the shift towards the usual theory would be much greater, however. This would accord with the opinion of virologists who have suggested that recombinants are derived in humans from a human component (A3) and an animal component (A2), with various preferences given to pigs and birds as the source of the animal component. It has been argued in support of this point of view that segments of human virus can be found that show considerable homology to segments of virus in animals. But if both are built from space-incident components the situation could hardly be otherwise.

If one were to take such a point of view, the main activator A1 would be space-incident, whereas the source of the virus would be terrestrial, and this would be an uneasy, implausible mixture. Almost inevitably, the next step would be to suppose A1 also to be of terrestrial origin, but this would lead back immediately to the epidemiologic difficulties set out above. Indeed, one would be returned more or less to the conventional theory with all its attendant problems.

Apart from the epidemiological difficulties, it would be hard to maintain global genetic uniformity in the influenza virus. The striking aspect of the genetic shifts that occur from decade to decade is that the shifts are everywhere synchronous to within a margin dictated by descent through the stratosphere, synchronous to within a time-scale of months to about a year. If A2 were contributed by a variety of animals, one might expect a thorough genetic mix-up, with pigs, horses, ducks, chickens, etc., all making their contributions contemporaneously in different geographical locations. Genetic uniformity forces near uniformity in incidence all over the Earth, which points strongly to an extra-terrestrial source for A2

Between 1860 and 1889 the death-rate in England and Wales died away almost to zero, and influenza was mild the world over, as seen for instance in Fig. A.3. Were pigs, horses, ducks and chickens everywhere suddenly inactive in their exudents of A2 over those decades? The question is only a detail, but it is representative of many details which cause trouble in an Earth-bound theory.

RELEVANT BIBLIOGRAPHY

At many places in the book we have made assertions based on detailed technical calculations that we were unable to repduce in full. Here we list a selection of our publications that not only shows how our ideas evolved over two decades of research, but also indicates where the reader can turn to for further justification of technical arguments.

A: Popular articles and books

1. 'Does epidemic disease come from space?', Fred Hoyle and Chandra Wickramasinghe, New Scientist, 17 November 1977

2. 'Influenza from space?', Fred Hoyle and Chandra Wickramasinghe, New Scientist, 28 September 1978

3. 'Where the microbes boldly went', Fred Hoyle and Chandra Wickramasinghe, New Scientist, 13 August 1981

4. 'Lifecloud: the origin of life in the galaxy', Fred Hoyle and Chandra Wickramasinghe (J.M. Dent Lond, 1978)

5. 'Diseases from Space', Fred Hoyle and Chandra Wickramasinghe (J.M. Dent Lond, 1979)

6. 'Space Travellers: the bringers of life', Fred Hoyle and Chandra Wickramasinghe (Univ.Coll.Cardiff Press, 1981)

7. 'Evolution from Space', Fred Hoyle and Chandra Wickramasinghe (J.M. Dent, Lond, 1981)

8. 'Is Life an Astronomical Phenomenon?', Chandra Wickramasinghe (Univ.Coll.Cardiff Press, 1982)

9. 'Cosmic Lifeforce', Fred Hoyle and Chandra Wickramasinghe (J.M Dent, Lond, 1988)

10. *Archaeopteryx* – The Primordial Bird: A case of fossil forgery, Fred Hoyle and Chandra Wickramasinghe (Christopher Davies, Swansea, 1986)

11. 'Our Place in the Cosmos', Fred Hoyle and Chandra Wickramasinghe (Weidenfeld & Nicholson, Lond, 1993)

B: Technical books and papers

1. 'On graphite particles as interstellar grains', F. Hoyle and N.C. Wickramasinghe, M.N.R.A.S., 124,417,1962

2. 'Interstellar Grains', N.C. Wickramasinghe (Chapman & Hall, Lond, 1967)

3. 'Interstellar grains', F.Hoyle and N.C. Wickramasinghe, Nature, 223, 459, 1969

4. 'Light Scattering Functions for Small Particles with Applications in Astronomy', N.C. Wickramasinghe (J. Wiley, NY, 1973)

5 'Formaldehyde polymers in interstellar space', N.C. Wickramasinghe, Nature, 252, 462,1974

6. 'Polyoxymethylene polymers as interstellar grains', N.C. Wickramasinghe, M.N.R.A.S., 170, 11P,1974

7. 'Formaldehyde polymers in comets', N.C. Wickramasinghe and V. Vanysek, Astrophys. Sp.Sc., 33, L19,1975

8. 'Composition of cometary dust: the case against silicates' D.A. Mendis and N.C. Wickramasinghe, Astrophys. Sp.Sci., 37, L13, 1975

9. 'Primitive grain clumps and organic compounds in carbonaceous chondrites', F. Hoyle and N.C. Wickramasinghe, Nature, 264, 45, 1976

10. 'Organic molecules in interstellar dust: a possible spectral signature at 2200A?', N.C. Wickramasinghe, F. Hoyle and K. Nandy, Astrophys. Sp.Sci., 47, L1, 1977

11. 'Polysaccharides and the infrared spectrum of OH26.5+0.6', F. Hoyle and N.C. Wickramasinghe, M.N.R.A.S., 181 51P,1977

12. 'Spectroscopic evidence for interstellar grain clumps in meteoritic inclusions', A. Sakata, N. Nakagawa, T. Iguchi, S. Isobe, M. Morimoto, F.Hoyle and N.C. Wickramasinghe, Nature, 266, 241, 1977

13. 'Pre-biotic molecules in Martian dust clouds', H.Abadi and N.C. Wickramasinghe, Nature, 267, 687, 1977

14. 'Polysaccharides and the infrared spectra of galactic sources', F. Hoyle and N.C. Wickramasinghe, Nature, 268, 610, 1977

15. 'Prebiotic polymers and infrared spectra of galactic sources', N.C. Wickramasinghe, F. Hoyle, J. Brooks, and G. Shaw, Nature, 269, 674, 1977

16. 'Identification of the 2200A interstellar absorption feature', F.Hoyle and N.C. Wickramasinghe, Nature, 270, 323, 1977

17. 'Origin and nature of carbonaceous material in the galaxy', F.Hoyle and N.C. Wickramasinghe, Nature, 270, 701,1977

18. 'Identification of interstellar polysaccharides and related hydrocarbons', F. Hoyle, N.C. Wickramasinghe and A.H. Olavesen, Nature, 271, 229,1978

19. 'Calculations of infrared fluxes from galactic sources for a polysaccharide grain model', F. Hoyle and N.C. Wickramasinghe, Astrophys. Sp.Sci.,53, 489,1978

20. 'Comets, ice ages and ecological catastrophes', F. Hoyle and N.C. Wickramasinghe, Astrophys. Sp.Sci., 53, 523,1978

21. 'Biochemical chromophores and the interstellar extinction at ultraviolet wavelengths', F. Hoyle and N.C. Wickramasinghe, Astrophys. Sp.Sci., 65, 241, 1979

22. 'On the nature of interstellar grains', F. Hoyle and N.C. Wickramasinghe, Astrophys. Sp.Sci., 66, 77,1979

23. 'Organic grains in space', F. Hoyle and N.C. Wickramasinghe, Astrophys. Sp.Sci., 69, 511, 1980

24. 'Organic material and the 1.5-4 micron spectra of galactic sources', F. Hoyle and N.C. Wickramasinghe, Astrophys. Sp.Sci., 72, 183,1980

25. 'Dry polysaccharides and the infrared spectrum of OH26.5+0.6' F. Hoyle and N.C. Wickramasinghe, Astrophys. Sp.Sci., 72, 247, 1980

26. 'Evidence for interstellar biochemicals', F.Hoyle and N.C. Wickramasinghe,

in Giant Molecular Clouds in the Galaxy, ed. P.M. Solomon and M.G. Edmunds, Pergamon, 1980

27. 'Is life an astronomical phenomenon?' C. Wickramasinghe, (University College, Cardiff Press, 1982)

28. 'Why Neo-Darwinism does not work', F.Hoyle and C. Wickramasinghe, (University College, Cardiff Press, 1982)

29. 'Comets – a vehicle for panspermia', F. Hoyle and N.C. Wickramasinghe, ed. C. Ponnamperuma, D. Reidel Publishing Co., 1981

30. 'Infrared spectroscopy of micro-organisms near 3.4 microns in relation to geology and astronomy', F. Hoyle, N.C. Wickramasinghe, S.Al-Mufti and A.H. Olavesen, Astrophys. Sp.Sci., 81, 489, 1982

31. 'Proofs that Life is Cosmic', Fred Hoyle and Chandra Wickramasinghe, Institute of Fundamental Studies, Sri Lanka, Memoirs, No.1. 1982

32. 'Infrared spectroscopy over the 2.9-3.9 micron waveband in biochemistry and astronomy', F. Hoyle, N.C. Wickramasinghe, S. Al-Mufti, A.H. Olavesen and D.T. Wickramasinghe, Astrophys. Sp.Sci., 83, 405-409, 1982

33. 'Organo-siliceous biomolecules and the infrared spectrum of the Trapezium nebula', F.Hoyle, N.C. Wickramasinghe and S. Al-Mufti, Astrophys. Sp.Sci.,86, 63,1982

34. 'The infrared spectrum of interstellar dust', F.Hoyle, N.C. Wickramasinghe and S. Al-Mufti, Astrophys. Sp.Sci.,86, 341,1982

35. On the optical properties of bacterial grains, I, N.L. Jabir, F.Hoyle and N.C. Wickramasinghe, Astrophys. Sp.Sci., 91, 327,1983

36. 'Organic grains in the Taurus interstellar clouds', F. Hoyle and N.C. Wickramasinghe, Nature, 305, 161,1983

37. 'Bacterial life in space', F.Hoyle and N.C. Wickramasinghe, Nature, 306, 1983

38. 'The spectroscopic identification of interstellar grains', F.Hoyle, N.C. Wickramasinghe and S. Al-Mufti, Astrophys. Sp.Sci., 98, 343,1984

39. '2.8-3.6 micron spectra of micro-organisms with varying H_2O ice content', F.Hoyle, N.C. Wickramasinghe and N.L. Jabir, Astrophys. Sp.Sci., 92, 439,1983

40. 'The extinction of starlight at wavelengths near 2200A', F.Hoyle, N.C. Wickramasinghe and N.L. Jabir, Astrophys. Sp.Sci.,92, 433, 1983

41. 'An object within a particle of extraterrestrial origin compared with an object of presumed terrestrial origin', F.Hoyle, N.C. Wickramasinghe and H.D. Pflug, Astrophys. Sp.Sci., 113, 209, 1985

42. 'Legionnaires' Disease: Seeking a wider cause', F.Hoyle, N.C. Wickramasinghe and J. Watkins, The Lancet, 25 May 1985, p.1216

43. 'The availability of phosphorous in the bacterial model of the interstellar grains', F.Hoyle and N.C. Wickramasinghe, Astrophys. Sp. Sci., 103, 189,1984
Wednesday 15th January 1997

44. 'From grains to bacteria', F.Hoyle and N.C. Wickramasinghe, University College, Cardiff Press, 1984

45. 'Living Comets', F.Hoyle and N.C. Wickramasinghe, University College,

Cardiff Press, 1985

46. 'Viruses from Space', F.Hoyle and N.C. Wickramasinghe, University College, Cardiff Press, 1986

47. 'On the nature of the interstellar grains', Q. Jl. R.A.S., 27, 21,1986

48. 'On the nature of the particles causing the 2200A peak in the extinction of starlight', F.Hoyle and N.C. Wickramasinghe, Astrophys. Sp. Sci.,122, 181,1986

49. 'The measurement of the absorption properties of dry micro-organisms and its relationship to astronomy', F.Hoyle, N.C. Wickramasinghe and S. Al-Mufti, Astrophys. Sp. Sci., 113,413, 1985

50. 'The viability with respect to temperature of micro-organisms incident on the Earth's atmosphere', F. Hoyle, N.C. Wickramasinghe and S. Al-Mufti, Earth, Moon and Planets, 35, 79,1986

51. 'Diatoms on Earth, Comets, Europa and in interstellar space', R.B. Hoover, F. Hoyle, N.C. Wickramasinghe, M.J. Hoover and S. Al-Mufti, Earth Moon and Planets, 35, 19,1986

52. 'The case for life as a cosmic phenomenon', F.Hoyle & N.C. Wickramasinghe, Nature, 322, 509, 1986

53. 'The case for interstellar micro-organisms', F.Hoyle, N.C.Wickramasinghe and S. Al-Mufti, Astrophys. Sp.Sci.,110, 401,1985

54. 'Evaporating grains in P/Halley's coma', M.K. Wallis, R. Rabilizirov and M.K. Wallis, Astron. Astrophys., 187, 801-806, 1987

55. 'Interstellar extinction by organic grain clumps', F. Hoyle and N.C. Wickramasinghe, Astrophys.Space.Sci., 140, 191, 1988

56. 'A diatom model of dust in the Trapezium nebula', Q. Majeed, N.C. Wickramasinghe, F. Hoyle and S. Al-Mufti, Astrophys. Space Sci., 140, 205, 1988

57. 'The organic nature of cometary grains', N.C. Wickramasinghe, F. Hoyle, M.K. Wallis and S. Al-Mufti, Earth, Moon and Planets, 40, 101, 1988

58. 'Mineral and Organic Particles in Astronomy', N.C. Wickramasinghe, F. Hoyle and Q. Majeed, Astrophys. Space Sci., 158, 335, 1989

59. 'Modelling the 5-30µm spectrum of Comet Halley', N.C. Wickramasinghe, M.K. Wallis and F. Hoyle, Earth, Moon and Planets 43, 145, 1988

60. 'Aromatic Hydrocarbons in very small interstellar grains', N.C. Wickramasinghe, F. Hoyle, and T. Al-Jubory, Astrophys. Space Sci., 158, 135, 1989

61. 'A unified model for the 3.28µm and the 2200A interstellar extinction feature', F. Hoyle and N.C. Wickramasinghe, Astrophys. Space Sci., 154, 143, 1989

62. 'Biologic versus abiotic models of cometary dust', M.K. Wallis, N.C. Wickramasinghe, F. Hoyle and R. Rabilizirov, Mon. Not. Roy.Astr.Soc., 238, 1165-1170, 1989

63. 'An integrated 2.5-12.5µm emission spectrum of naturally occurring aromatic molecules', N.C. Wickramasinghe, F. Hoyle and T. Al-Jubory, Astrophys. Space Sci., 166, 333, 1990

64. 'Sunspots and influenza', F. Hoyle & N.C. Wickramasinghe, Nature, 343, 304, 1990

65. 'Influenza – evidence against contagion: discussion paper', F. Hoyle & N.C. Wickramasinghe, J.Roy.Soc.Med., 83, 258, 1990

66. 'The Theory of Cosmic Grains', Fred Hoyle and Chandra Wickramasinghe (Kluwer Academic Press, 1990)

67. 'The implications of life as a cosmic phenomenon: The anthropic context', F. Hoyle and N.C. Wickramasinghe, J.Brit.Interplan.Soc., 44,77-86,1991

68. 'Cometary habitats for primitive life', M.K. Wallis, N.C. Wickramasinghe and F.Hoyle, Adv.Space Res., Vol.12, No.4, pp(4)281-285, 199

69 'Triton's eruptions analogous to Comet Halley's?', M.K. Wallis and N.C. Wickramasinghe, Adv.Space Res., Vol 12, No.11, pp133-138, 1992

70. 'Comets as a source of interplanetary and interstellar grains', F. Hoyle and N.C. Wickramasinghe, in Origin and Evolution of Interplanetary Dust (eds. A.C.Levasseur-Regourd and H. Hasegawa), 235-240, Kluwer Academic Publishers, 1991

71. 'Comet Halley's remote outburst', M.K. Wallis and N.C. Wickramasinghe, The Observatory, 112, 228-234, 1992

72. 'Cosmic Grains', N.C. Wickramasinghe in 'Infrared Astronomy' (eds. A. Mampsaso, M. Prieto & F. Sanchez) Proceedings of the 4th Canary Islands Winter School of Astrophysics (Cambridge University Press) (ISBN 0521 464625) pp 275-299, 1994

73. 'The cometary hypothesis of K/T mass-extinctions', N.C. Wickramasinghe and M.K. Wallis, Monthly Notices of the Roy. Astr. Soc., 270, 420-426, 1994

74. 'Extinction of Dinosaurs: a possible novel cause', S. Ramadurai, D. Lloyd, M. Wallis & N.C. Wickramasinghe. Symposium F3.1 World Space Congress, Washington, 'Extraterrestrial Organic Chemistry and the Origins of Life', Adv. Space Res., 1994

75. 'Critique of Fischer-Tropsch type reactions in the solar nebula', S. Ramadurai, F. Hoyle and N.C. Wickramasinghe, Bull.Astr.Soc. India 21, 329-334, 1993

76. 'Role of major terrestrial cratering events in dispersing life in the solar system', M.K. Wallis & N.C. Wickramasinghe, Earth and Planetary Science Letters, 130, 69-73, 1995

77. 'Biofluorescence and the extended red emission in astrophysical sources', F. Hoyle and N.C. Wickramasinghe, Astrophys. Sp.Sc., 235, 343-347, 1996

78. 'Very small dust grains (VSDP's) in Comet C/1996 B2 (Hyakutake)', N.C. Wickramasinghe and F. Hoyle, Astrophys.Sp.Sc., 239, 121-123, 1996

79. 'Eruptions from comet Hale-Bopp at 6.5AU', N.C. Wickramasinghe, F. Hoyle and D. Lloyd, Astrophys.Sp.Sc., 240, 161-165, 1996

80. 'Infrared signatures of prebiology – or biology', N.C. Wickramasinghe, F. Hoyle, S. Al-Mufti; and D.H. Wallis in Astronomical and Biochemical Origins and the Search for Life in the Universe, e.d. C.B. Cosmovici, S. Bowyer and D. Werthimer (Editrice Compositori, 1997)

INDEX